Solarities

Forerunners: Ideas First

Short books of thought-in-process scholarship, where intense analysis, questioning, and speculation take the lead

FROM THE UNIVERSITY OF MINNESOTA PRESS

After Oil Collective; Ayesha Vemuri and Darin Barney, Editors
Solarities: Seeking Energy Justice

Arnaud Gerspacher
The Owls Are Not What They Seem: Artist as Ethologist

Tyson E. Lewis and Peter B. Hyland
Studious Drift: Movements and Protocols for a Postdigital Education

Mick Smith and Jason Young
Does the Earth Care? Indifference, Providence, and Provisional Ecology

Caterina Albano
Out of Breath: Vulnerability of Air in Contemporary Art

Gregg Lambert
The World Is Gone: Philosophy in Light of the Pandemic

Grant Farred
Only a Black Athlete Can Save Us Now

Anna Watkins Fisher
Safety Orange

Heather Warren-Crow and Andrea Jonsson
Young-Girls in Echoland: #Theorizing Tiqqun

Joshua Schuster and Derek Woods
Calamity Theory: Three Critiques of Existential Risk

Daniel Bertrand Monk and Andrew Herscher
The Global Shelter Imaginary: IKEA Humanitarianism and Rightless Relief

Catherine Liu
Virtue Hoarders: The Case against the Professional Managerial Class

Christopher Schaberg
Grounded: Perpetual Flight . . . and Then the Pandemic

Marquis Bey
The Problem of the Negro as a Problem for Gender

Cristina Beltrán
Cruelty as Citizenship: How Migrant Suffering Sustains White Democracy

Hil Malatino
Trans Care

Sarah Juliet Lauro
Kill the Overseer! The Gamification of Slave Resistance

Alexis L. Boylan, Anna Mae Duane, Michael Gill, and Barbara Gurr
Furious Feminisms: Alternate Routes on *Mad Max: Fury Road*

Ian G. R. Shaw and Marv Waterstone
Wageless Life: A Manifesto for a Future beyond Capitalism

Claudia Milian
LatinX

Aaron Jaffe
Spoiler Alert: A Critical Guide

Don Ihde
Medical Technics

Jonathan Beecher Field
Town Hall Meetings and the Death of Deliberation

Jennifer Gabrys
How to Do Things with Sensors

Naa Oyo A. Kwate
Burgers in Blackface: Anti-Black Restaurants Then and Now

Arne De Boever
Against Aesthetic Exceptionalism

Steve Mentz
Break Up the Anthropocene

John Protevi
Edges of the State

Matthew J. Wolf-Meyer
Theory for the World to Come: Speculative Fiction and Apocalyptic Anthropology

Nicholas Tampio
Learning versus the Common Core

Kathryn Yusoff
A Billion Black Anthropocenes or None

(Continued on page 77)

Solarities

Seeking Energy Justice

After Oil Collective

Ayesha Vemuri and Darin Barney, Editors

University of Minnesota Press

MINNEAPOLIS

LONDON

ISBN 978-1-5179-1414-1(PB)
ISBN 978-1-4529-6811-7 (Ebook)
ISBN 978-1-4529-6868-1 (Manifold)

Published by the University of Minnesota Press, 2022
111 Third Avenue South, Suite 290
Minneapolis, MN 55401-2520
http://www.upress.umn.edu

Available as a Manifold edition at manifold.umn.edu

After Oil Collective

The After Oil Collective is a subgroup of the Petrocultures Research Group that meets periodically for collaborative work. *Solarities: Seeking Energy Justice* is one of the outcomes of this work.

Aaron Kirkey, Amelia Moore, Ashley Dawson, Ayesha Vemuri, Bob Johnson, Brent Ryan Bellamy, Burç Köstem, Caleb Wellum, Claire Ravenscroft, Cymene Howe, Darin Barney, Derek Gladwin, Dominic Boyer, Elizabeth Miller, Eva-Lynn Jagoe, Gökçe Günel, Graeme Macdonald, Gretchen Bakke, Hannah Tollefson, Heather Davis, Ian Clarke, Imre Szeman, Jamie Cross, Jeff Diamanti, jenni matchett, Jennifer Wenzel, Jessie L. Beier, Joel Auerbach, Jordan B. Kinder, Josefin Wangel, Kim Förster, Kyle Whyte, Laurence Miall, Lev Bratishenko, Louis Beaumier, Maize Longboat, Mark Simpson, Mette M. High, Mirra-Margarita Ianeva, Nandita Badami, Nicole Starosielski, Rachel Webb Jekanowski, Rafico Ruiz, Rhys Williams, Shane Gunster, Sheena Wilson, Shirley Roburn, Simon Orpana, Stacey Balkan, Tomas Ariztía, and Yuriko Furuhata.

Solarity (sō′lərĭ-tē): n. a state, condition, or quality developed in relation to the sun, or to energy derived from the sun.

Contents

Situating Solarity

SOLARITY IS THE CONDITION we inhabit as we struggle to make worlds and build futures out of the ruins of petrocapitalism. It inspires hope, creativity, and joy, even as it carries the weight of our guilt, shame, and anger at what we—or, more precisely, some of us—have been and done. It is worked into nearly everything, transforming matter into blooms and desiccations; it is existential and infinitesimal all at once. Solarity is, fundamentally, a relational condition; the heat, light, and gravity of the sun are all so deeply intertwined with earthly life that there is no "outside" of solarity. As we turn, in this moment of planetary crisis, toward the sun, we seem to be seeking both a new energy regime and a new set of human and more-than-human relations. It is also an inherently ambivalent condition. The chapters that follow are the combined work of more than fifty contributors. As a result, the meanings attached to solarity—an emergent concept being shaped by multiple voices—will be many. This is a feature, not a bug (though bugs are generative too). Solarity offers an illuminating yet blinding array of new (and ancient) histories, stories, mythologies, promises, materials, and relations that are deeply ambiguous: neither "good" nor "bad," neither wholly "positive" nor "negative," neither "bright" nor "bleak." Some solarities could be revolutionary. Others will not be. These qualifiers and possibilities are bound to the particular circumstances from which they arise—historical and otherwise—as well as the subjective orientations of those experiencing them. They are, in other words, *situated.*

Although the condition of solarity has always been central to earthly cycles of life and death, the current ecological crisis invites a new orientation to the sun focused primarily on solarity as *energy* or, more accurately, as *fuel.* Seen this way, solarity is also an excessive flood of energy abundance. The sun's energy, especially in discourses of solar energy and solar transitions, is depicted in grand, overflowing terms. Think of these refrains that have become truisms in the solar imaginary: "more energy falls on the world's deserts in six hours than the total global energy consumption in a year" or "if we took all of the sun's energy from everywhere on the planet, in one hour, we would have enough energy to power the world for an entire year." Solar abundance also turns to the moon. Some have proposed a Luna Ring,[1] with electric power to be generated through a belt of solar cells surrounding the eleven-thousand-kilometer lunar equator. Power will be beamed to Earth from the near side of the moon with the Luna Ring existing in continuous sunlight, thereby doubling the power generated on Earth in the same twenty-four-hour period.

Although claims about the quantity of energy that can be seized from the sun may be accurate, they also tend toward the consumptive. They channel the human user of energy into a web of solarity that is about capture and utility. Solarity can ebb into the domain of gluttony when there is ever, forever, more-than-enough. The power of the sun is made to figure as providing for human lives, all calculated through consumption. A seemingly insatiable human desire for power pairs well with solarity's plenitude. Though there is enough power for all in solarity's promise, theoretically, it is also true that such abundance rarely gets distributed evenly. But perhaps solarity can change that.

1. Solar System Exploration Research Virtual Institute, "The Luna Ring Concept," https://sservi.nasa.gov/articles/the-luna-ring-concept/.

The promise of solarity's redistribution must contend not only with the temporal and spatial logics of the sun's radiation but also with the forms of technological and infrastructural mediation required to convert solar energy into electrical energy. It also requires us to overcome decades of negative campaigning by the many-tentacled arms of the fossil fuel complex, from media to politics, that have not only consistently emphasized—falsely and heinously—its material limits but also preposterously claimed it as destructive to industry, environment, and even health.[2] Though solarity promises a transcendence of limits in terms of energy, the limits of space and materials on Earth remain. Solarity, then, requires a reckoning with our material and affective attachments to the present. For those of us in the overdeveloped world living lives of hyperconsumption, shame, guilt, and anger are often our first emotional responses to the climate crisis. Solarity asks for much more: it demands a radical hope as well as a radical reassessment of what, of the present, we might wish to retain as we attempt to rescue our future.

2. See, e.g., Michael Mann, *The New Climate War: The Fight to Take Back Our Planet* (New York: Public Affairs/Hatchette, 2021).

The Promise of Solarity

THE TURN TOWARD THE SUN, toward embracing solarity, is replete with desire and hope. Hope for a future that grasps the sun's abundance without the need for mediation. Hope for a future when "civilization's" reliance on the buried sunlight of fossil fuels is overcome. Hope for a more egalitarian and just future for human and nonhuman relations. Hope that we will overcome the devastating spiral of planetary destruction, death, and extinction that appears on the horizon. Hope that we can continue living as we do in the overdeveloped West by turning toward solar fuel. Hopes that cascade and multiply and are awash with both harmonies and dissonances. Though many share the vision of a just solar future, delving deeper into specific solar desires reveals that there is not, and cannot be, one Solarity; rather, there is a proliferation of plural solarities that manifest a spectrum of desires.

Solarities are orientations to the energy of the sun. These orientations exceed human relations to the sun's rays to also include relationships with lands, minerals, waters, animals, and people, relationships that are mediated by and materialized in infrastructure. What will the infrastructures of these solarities be? What will they be made of, and what will they make of us? One of the desires attending solar energy is that, in its universal availability and infinite supply, it does not require the sorts of mediation necessary for energy to become fuel. Yet this is imaginary: an orientation that approaches the sun as a source of unlimited energy, capable of saving the planet by replacing fossil fuels, necessarily implies infrastructural mediation with specific material implications.

When we discuss solarity, it often seems to imply solar energy tout court. But we would rather orient ourselves to solarity as naming a way of collective being that is consonant with the generous outpour of energy from the sun. Solarity uncovers a collective desire to share and distribute energy as a way of regulating the social order. But today, this desire cannot be disentangled from an impetus to overturn the prevailing energy order and its coextensive injustices. Petroculture is premediating the way solarity unfolds. The utopian aspirations of solarity will be informed by global oil's history of social and environmental exploitation. In other words, while solarity might supplant the petrocultural regime, it is also at risk of succumbing to its slow violence and necropolitics.[1] What seems certain is that the transition to solarity will be messy, incomplete, and unevenly distributed.

For some, solar names the promise of clean energy; it is also the promise of *infinite* energy. This is due to the sheer amount of energy produced by the sun. And to add to the good news, there's no need to worry about "peak solar" in the way that some have fretted about "peak oil": we can count ourselves safe for the next five billion years, until the sun begins to transition into a red giant. When we commonly speak about "energy," what we are really referring to is fuel: matter that can be made to release energy. Every form of fuel we currently use demands the production of physical infrastructures to create energy (from fireplaces to nuclear power stations); in the process, as fuel becomes energy, it always leaves a physical trace. With solar power, we believe we have found a way to cut fuel out of the picture of energy production. The dream of solar is that we can access energy as energy: energy without mediation, energy without

1. See Rob Nixon, *Slow Violence and the Environmentalism of the Poor* (Cambridge, Mass.: Harvard University Press, 2013); Achille Mbembe, "Necropolitics," *Public Culture* 15, no. 1 (2003): 11–40.

the need for fuel and so without leaving any trace of its use. This is the dream of infinite energy without needing to worry about its impact, either as extraction or as emission.

Thus does solar power present itself as the solution, but already, we have a problem. Who is this "we" that "commonly speaks about energy" in this way? The pretense of the first person plural is exposed the moment energy enters the picture. Some relate to energy in this way; many others do not, or would not if they were not forced to by those who do. Some (mostly men, mostly white, mostly Anglo/Euro/Christian descendant, mostly rich, mostly on land that is not their own) accumulate and enjoy the benefits of fuel and the energy it generates, whereas others (mostly women, mostly racialized, mostly from "Other" traditions, mostly poor, mostly Indigenous and dispossessed) have themselves historically and contemporarily been treated as fuel, a source of energy to be extracted, expended, and exhausted for the sake of someone else's good life. Those who inhabit what Macarena Gómez-Barris calls the "extractive zone"[2] (of which there are many) have a very different relationship to energy than "we" do. It might be true that we have all always been solar, in the sense that people everywhere have lived by harvesting the energy of the sun in one way or another, but it would be a mistake to believe that this means we have all been or will be solar in the same way or in the ways we would choose. There has been and will be no solarity, only solarities, and the diverse characters of these solarities will be determined by the relations, not the source, of their energy.

Solar thus contains a double promise: energy without fuel and an infinite amount of energy. In her essay on the discourse of "zero"—"zero carbon, zero waste, zero landfill, zero emissions, (net) zero energy"[3]—in the context of large-scale solar projects in India,

2. Macarena Gómez-Barris, *The Extractive Zone: Social Ecologies and Decolonial Perspectives* (Durham, N.C.: Duke University Press, 2017).

3. Nandita Badami, "Counting on Zero: Imaginaries of Energy and Waste in the New Green Economy," *Platypus: The Castac Blog* (blog), October 21, 2016, http://blog.castac.org/2016/10/counting-on-zero/.

Simon Orpana, *Solar Surfer,* 2021. Pen and ink.

Nandita Badami writes, "Zero performs effective material and ideological work precisely because it simultaneously indexes both nothingness and infinitude." Solarity as a net zero condition reproduces the fantasy of using energy without environmental consequences. No fuel means no spent fuel rods to bury, no carbon dioxide to manage, no flooded valleys from hydro projects to ameliorate, no torn-apart and poisoned land to recondition. In the drama called "sustainability," solar plays the part of the hero who appears in the nick of time to save us from ourselves. Solar stands over the dead body of fossil fuels, sword raised to the sun, leading us forward into a future in which energy is energy and fuel is left for history books.

This version of solar as an energy source immediately available, infinite in supply and negligible in cost, settles upon the sun as but the latest frontier of an extractive enterprise whose names have been slavery, colonialism, industrial capitalism, imperialism. Among its many potentials, solarity could be the new name for this enterprise. But this seems counterintuitive. How can one possess the sun and own what is infinite? We do well to recall that the histories and present of extractive enterprise have always and everywhere involved projecting the possibility of property, with its attendant relations, upon people, places, and things previously conceived as common and infinite.[4] This reminds us that, in many places, the projection of commonness that precedes appropriation has itself been an act of dispossession and part of the ongoing legacy of the colonial, including settler and neocolonial—formations in which many of us live.

What happens to property in a world awash in energy? Moreover, what happens to a world awash in yet another type of energy that has been forcibly converted into property? One possibility is that solarity might be continuous with the capitalist, masculinist, racist, colonialist, and imperialist extractive enterprises that have defined the fossil fuel era globally. Still, the energy promised by solar can't help but lead us to speculate about how else people might live once they have access to infinite, clean energy. The possibilities of lives transformed by ready access to solar energy have been expressed with particular hope and force in relation to geographies historically consigned to energy poverty by the colonialisms and imperialisms of the Anglo-Euro-American Global North.[5] Could solarity mate-

4. Silvia Federici, *Caliban and the Witch: Women, the Body, and Primitive Accumulation* (New York: Autonomedia, 2014).

5. Bill McKibben, "The Race to Solar-Power Africa," *New Yorker,* June 19, 2017, http://www.newyorker.com/magazine/2017/06/26/the-race-to-solar-power-africa; Jamie Cross, "The Solar Good: Energy Ethics in Poor Markets," *Journal of the Royal Anthropological Institute* 25, no. S1 (2019): 47–66.

rialize different ways of being in relation to one another and to the plurality of nonhuman others with whom our fates are entangled?[6]

Answers to this question will depend heavily on whether and how people approach the promise of solar power as either infinite or not. Until very recently, the upper and middle classes of the wealthy capitalist countries have *always* used energy as if it were infinite, worrying little about the repercussions of the fuels they've used or the uses to which their fuels are put. Peak oil temporarily gave those of us in this situation pause, but it is global warming that has caused us to reflect on the processes and practices by which we transform the energy of the sun into fuels we can use, and that has caused us to think more seriously about the implications of using these fuels as if they were infinite. What will solarity mean if a primary use of the sun as fuel is to power air conditioners that make it possible (for some) to withstand the sun's relentless heat? The desire for infinite and limitless energy seems, in many ways, too deeply intertwined with capitalist obsessions with limitless growth that have ignored the needs of the many in favor of serving the greed of the few. Can the desire for infinite energy be disentangled from these extractive and oppressive histories of unfettered growth? Or does a more solidarity-oriented solarity require a new vocabulary of imagination and desire that prioritizes subsistence and satiation over the ever-receding horizon of infinitude?

Ursula Le Guin has written that the "utopian imagination is trapped . . . in a one-way future consisting only of growth."[7]

6. Kim TallBear, "Theorizing Queer Inhumanisms: An Indigenous Reflection on Working beyond the Human/Not Human," *GLQ: A Journal of Lesbian and Gay Studies* 21, no. 2–3 (2015): 230–35; Marisol de la Cadena, *Earth Beings: Ecologies of Practice across Andean Worlds* (Durham, N.C.: Duke University Press, 2015); Donna Haraway, *Staying with the Trouble: Making Kin in the Chthulucene* (Durham, N.C.: Duke University Press, 2016); Eduardo Kohn, *How Forests Think: Toward an Anthropology beyond the Human* (Berkeley: University of California Press, 2013).

7. Ursula K. Le Guin, *Dancing at the Edge of the World: Thoughts on Words, Women, Places* (New York: Grove Press, 1997).

SOLAR, FARMS: ON PLANNING FOR SOLARITY

Imre Szeman

Renewable energy systems can be hard to set up. Three words sum up some of the unexpected limits that need to be addressed: *property, zoning,* and *bumpy.*

PROPERTY: There is nothing compelling any private landowner to make her property available for renewable energy systems. In Canada, close to 90 percent of all land is publicly owned, which might seem to offer a solution. But there are two hiccups to this. The first is that the percentage of Crown land drops significantly as one moves closer to Canada's major population centers, where most of the renewable energy would be needed. The second hiccup? The Canadian government is not in the energy business.

ZONING: Whether private or public, not all property is made alike. Decisions have been made about what can go where—housing, businesses, and industry. There are wetlands and watersheds, natural and recreation parks, conservation lands and farmlands. None of these can house renewable energy systems.

BUMPY: Earth isn't flat. Even if the equation of property + zoning seems to be pointing to a positive outcome for renewables, slope, drainage, and soil composition mean that, in many cases, the outcome is zero.

Ecomodernist accounts of energy transition live in the ruins of such an imaginary, even as the misery and violence required by endless "growth" become increasingly undeniable—solar panels on green pastures, Reagan's shining city in the background, shimmering blue with their reflection. What unites such fantastical narratives of technology and growth and the more fatalist responses to ecological disaster that seek to "administer" the catastrophe is their investment in scarcity as the definitive ecological problem. In the ruins of growth, scarcity is either overcome by technology or contained by force, borders, and cruelty.

Perhaps solarity requires of us another imaginary, one shared by many cultures—not of growth but of abundance. Abundance is what literally moves with a wave, *ab + undare,* of undulation. Hence it is nothing like the overcoming of limits, the stockpiling of surplus siphoned to increase production, but on the contrary requires an embracing of earthly metabolisms, even of degrowth. The ethos of abundance exists all around us, in insisting that there is always more room at the table, in practices of obligatory sharing, expenditure, and the commons. Through it, one may hope to vacate the very desire for accumulation that drives economic expansion. Yet abundance might also be the name of a political strategy for embracing collective limits, of warding off growth sometimes through equitable redistribution, other times through glorious expenditure. An abundant energy transition would untether our visions of the good life from narratives of growth and private accumulation and mobilize behind equitable sharing of energy as a collective good.

In *The Accursed Share,* Georges Bataille[8] proposes a theory of the economy that begins with the sun and the energy that it produces. "Solar energy," Bataille writes, "is the source of life's exuberant development. The origin and essence of our wealth are given in the radiation of the sun, which dispenses energy—wealth—without any return. The sun gives without ever receiving."[9] This originary dispensation is key to what Bataille describes as a general economy. The general economy is constituted by expenditure and squandering, because the energy of the sun is always in excess, impossible to contain and control. By contrast, the human economy is constituted as a restricted one that operates as if there were a scarcity of energy and other resources and so is organized around the control and management of them. As Amanda Boetzkes[10] writes, "capitalism's failure to acknowledge our innate solarity, and its fundamental prohibition of expenditure, results in the extreme pressure to accumulate energy without waste (in the form of profit) and a collective

8. Georges Bataille, *The Accursed Share,* vol. 1, trans. Robert Hurley (New York: Zone Books, 1987).

9. Bataille, 1:28.

10. Amanda Boetzkes, "Solar," in *Fueling Culture: 101 Words for Energy and Environment,* ed. Imre Szeman, Jennifer Wenzel, and Patricia Yaeger (New York: Fordham University Press, 2017), 315.

drive toward planetary destruction." Here solarity serves as the countermodel for social relations based on generous expenditure rather than hoarding. Bataille's solarity is a fundamentally *redistributive* condition—to cite the example Bataille offers: "a transfer of American wealth to India without reciprocation."[11] Thus, Bataille observes, "changing from the perspective of *restrictive* economy to those of a *general* economy actually accomplishes a Copernican transformation: a reversal of thinking—and of ethics."[12]

Something like the Copernican "reversal of thinking" attributed by Bataille to solarity has animated many designs for large-scale social transformations based on the specific materiality of the sun. Boetzkes describes the vision of a decentralized, ecological community advanced by Murray Bookchin's postscarcity anarchism as predicated on a belief in solar power as "an inexhaustible source of energy, freely and equally available."[13] David Schwartzman has long insisted that we are approaching the possibility of "solar communism,"[14] which Schwartzman describes as "a global civilization realizing Marx's aphoristic definition of communism for the twenty-first century: 'from each according to her ability, to each according to her needs,' referring to both humans and ecosystems." Here the advent of solar solves two big problems at once, positioning human beings in a better relation to nature and to one another. In Schwartzman's view of things, solar does away with the rationale for and support of the military–industrial complex; the (virtually) free energy of solar also does away with scarcity and with economic value as we currently understand it.

Similarly, Hermann Scheer, once a German parliamentarian and architect of that country's famed *Energiewende* (energy transition), declared, "Making the ground-breaking transition to an economy

11. Bataille, *Accursed Share,* 1:40.
12. Bataille, 1:25.
13. Boetzkes, "Solar," 316.
14. David Schwartzman, "Beyond Eco-catastrophism: The Conditions for Solar Communism," *Socialist Register* 53: 143–60.

based on solar energy will do more to safeguard our common future than any other economic development since the Industrial Revolution."[15] For Scheer, this was because the particular material properties of solar energy and its infrastructures promised to "reestablish the links between the development of the economy as whole and environmental cycles, stable regional business structures, cultures and democratic institutions," a possibility the fossil fuel economy has all but obliterated. As Dominic Boyer observes, in this respect, Scheer seeks "a revolutionary leap forward into a new form of sociality, one that is energy intensive and technologically enabled but resolutely local, sustainable and diverse."[16]

When we think about solar, we need to be alert to its ideological function, which is yet again to erase social relations, interspecies relations, material relations, and finitude from the picture of energy use. As Badami observes, the laudable goal of reaching zero through solar and other renewable infrastructures can leave uninterrogated the paradigms of technological and economic growth reproduced by such plans. Badami writes, "The ideological and material work that the zero seems, ultimately, to perform is to maintain the fiction that we can have our cake and eat it too. It lets us constantly consume, and then calibrate—in order to 'lighten the footprint'—and allows us the comfort of not having to reimagine the potential limitlessness of consumption."[17]

Writing about the accelerating arrival of photovoltaic technologies to regions of energy poverty in South Asia and sub-Saharan Africa, Jamie Cross points out that the benefits of off-grid solar electrification for the energy impoverished exist alongside the opportunities for capital intensification that these same projects offer to the already enriched. As Cross puts it, "for many management

15. Hermann Scheer, *The Solar Economy* (London: Earthscan, 2002), 33.
16. Dominic Boyer, "Infrastructure, Potential Energy, Revolution," in *The Promise of Infrastructure,* ed. Nikhil Anand, Akhil Gupta, and Hannah Appel (Durham, N.C.: Duke University Press, 2018), 237.
17. Badami, "Counting on Zero."

and business executives in off-grid solar companies, selling solar power to people living in chronic energy poverty presents itself as an ethical-economic utopia: the opportunity to express care for others and the environment at the same time as fulfilling a fiduciary duty of care to investors and shareholders."[18] To say that solar promises infinite, clean energy might well be to say that it allows us to continue to think of energy in the same way that many of us always have, even as the dream of energy sufficiency provides an alibi for erasing the political, ethical, and economic relationships that subtend how energy becomes power.

We can detect this bright depiction of "solar power" in the popular image bank of solarity. A quick internet search for "solar power," for example, presents standard images of large-scale farms and arrays tending toward an aesthetics of gleaming futurity and industrial-scale possibility. The future is bright and upscaled in such visions. Cross's warning, however, of the reality of "green dystopias" has powerful representative examples in some solar imaginaries that, although not downplaying the potential power of the sun as a formidable force in our energy futures, suggest a possible dystopian turn in solarity, if the kinds of relations we outline herein are not successfully realized. Perhaps the most resonant recent example is the shocking opening sequence of Denis Villeneuve's *Blade Runner 2049*, involving a flyover of a future California where a monochrome and smoggy landscape stocked with solar panels fills the screen to project a darkly ecological mood of entropy, violence, and decay. The solar horizon here is not a sunny one, and we are immediately asked to confront our expectations of solarculture's set associations. The promise of solarity, then, is always ambivalent.

18. Cross, "Solar Good," 2.

Solar Materialisms

SOLARITY IS A MATERIAL PROSPECT that invites a materialist response—but, which materialism? There is the historical materialism we have inherited from the Marxist tradition, with its attention to structural transformations in the forces, relations, and mode of production; class exploitation; antagonism and struggle; and the ideological and subjective forms and practices attached to these. And then there are the so-called new materialisms, brought to us by thinkers and practitioners arrayed across feminist science and technology studies, Indigenous place-thought and practice, postcolonial and Black anthropology, geography and fiction, and the ontological turn in European philosophy. In various and distinctive ways, each of these calls our attention to material relations with and between the nonhuman beings formerly known and treated as objects, beasts, resources, environments, and machines. These materialisms are often positioned as opposed, but it seems clear that comprehending solarity requires both of them, historical and new materialisms proceeding not in parallel but roped together in a braid.

For example, capitalism conditions us to regard the sun as a source of abundant energy that, in turn, frames our relations with other, Earth-bound beings and materials as one of limitation. In this view, solarity is frustrated by material limits on storing the sun's energy as fuel. This poses a problem for fueling the future in the same ways as the present, ways that privilege capital at the expense of human and more-than-human beings and environments. Perhaps this is one way of explaining why such a widespread harnessing of solar energy seems for some a faraway dream. At the same time, this current material reality signals the ways in which solarity might resist the legacies of extractive settler colonialism and capitalist modes of accumulation bound up with life in the fossil economy. On the other hand, despite the potential for environmentally clean energy on a large scale, solar economies may nevertheless be haunted by the history of colonial extractivism, labor exploitation, and capital accumulation bequeathed to them by the fossil fuel era. Sorting through the diverse material possibilities of solarity will require multiple materialisms, not just one.

A persistent strain of petrocultural studies insists that the culture and politics of societies energized by carbon-based fuels are not just incidentally related to the distinctive material properties of coal and oil.[1] Infrastructure matters, but so does, well, *matter*. There is an environmentalism in which deploying solar energy to sustain existing social and economic relations while slowing global warming by reducing greenhouse gas emissions associated with fossil fuels is *good enough*.[2] Solarity holds out the promise of more. This is partly because transitions are always periods of uncertainty, overdetermination, and potential. Things can be different. It is also because the sun in the sky is different from the oil in the ground in ways that suggest a range of alternative possibilities for organizing our relationship to it and to each other, possibilities that differ substantially from how our relationships to fossil fuels have been

1. Stephanie LeMenager, *Living Oil: Petroleum Culture in the American Century* (Oxford: Oxford University Press, 2014); Timothy Mitchell, *Carbon Democracy: Political Power in the Age of Oil* (London: Verso, 2011); Sheena Wilson, Adam Carlson, and Imre Szeman, eds., *Petrocultures: Oil, Politics, Culture* (Montreal: McGill-Queen's University Press, 2017).

2. Varun Sivaram, *Taming the Sun: Innovations to Harness Solar Energy and Power the Planet* (Cambridge, Mass.: MIT Press, 2018).

arranged. It is not without reason that the potential of the sun to energize fundamentally transformed economic, political, and social relations has captured the imagination of a range of otherwise quite divergent thinkers.

Writing about solar communism, David Schwartzman sees the promise of initiating a transition from a bleak capitalism to a brighter, more communal future organized by the energy of the sun. For Andreas Malm, the prospects of a solar future amount to a return to the "flow"—a return to harnessing inexhaustible energy sources that necessitate a reconfiguration of the rhythms of everyday life in a way attuned to that flow, rather than to the rhythms of capitalist production.[3]

Does solarity by its very nature offer a brighter future? Is solar energy necessarily communist, and communism necessarily solar? Brightness is how we perceive the sun, but the layers of mediation necessary to harness its abundance create conditions for replicating the same uneven bleakness found in the fossil economy. It is here that materials reappear to temper the solar imaginary. Mining the sun for its energy requires solar photovoltaic systems (PV) and batteries to store the energy they generate. It requires inverters for converting the direct current electricity produced by PV systems to the alternating current common to most appliances. At scale, it also requires advanced computing and data processing technologies to manage the grids into which solar-generated electricity flows. The process of creating and operating PV systems is materials and energy intensive. It involves the use of poisonous and toxic chemicals, including cadmium compounds, hexafluoroethane, silicon tetrachloride, and lead, the life cycles of which exceed the anthropocentric circuit of extraction, manufacture, use, and disposal. Lithium ion batteries are the ones most commonly used in PV systems. There are a range of issues associated with using lithium, including the

3. Andreas Malm, "A Return to the Flow? Obstacles to the Transition," in *Fossil Capital*, 367–88 (London: Verso, 2016).

amount of water required in its mining process (half a million gallons per metric ton of lithium), the generation of toxins in the process of lithium processing, and the colonial displacements that nearly always accompany its extraction. South America's Lithium Triangle (a region that includes Argentina, Bolivia, and Chile) is estimated to hold more than half of the world's supply. Whether this fact places these countries, all former colonial and existing neocolonial extractive zones, in a position of strategic advantage or intensified environmental, economic, and social precarity is a historical question. We do know that Latin America is a region in which the burdens of mining (and resistance to its injustices) have fallen heavily on women, Indigenous peoples, and workers, while its benefits have flowed and accumulated elsewhere.[4]

The environmental and political implications of the large-scale extraction, processing, and disposal of other elements involved in battery production—cobalt and nickel—are as troubling as all the others listed here (and this is far from a complete list). Solar power also has implications for land and water use. Depending on the system in use—utility-scale PV systems or concentrated solar thermal power (CSP) facilities—3.5–16.5 acres are required per megawatt generated. Even if we believe the energy of the sun to be limitless, the availability of land is not, and land used for solar is land that can't be used for other purposes. CSP plants also need water for cooling and cleaning, as dust, dirt, bird droppings, and other organic matter greatly reduce the efficiency of solar panels. These facilities are often located in areas with dry (i.e., sunny) climates, adding an additional and lucrative source of demand upon already limited

4. Maristella Svampa, "Commodities Consensus: Neoextractivism and Enclosure of the Commons in Latin America," *South Atlantic Quarterly* 114, no. 1 (2015): 65–82; Katy Jenkins, "Women Anti-mining Activists' Narratives of Everyday Resistance in the Andes: Staying Put and Carrying On in Peru and Ecuador," *Gender, Place, and Culture* 24, no. 10 (2017): 1441–59; Kuntala Lahiri-Dutt, "Digging Women: Towards a New Agenda for Feminist Critiques of Mining," *Gender, Place, and Culture* 19, no. 2 (2012): 193–212.

SABOTAGING SOLAR POWER IN CHALISGAON
Amanda Boetzkes

In February 2018, a YouTube video showing solar power plant workers destroying solar panels with hammers and sticks in Chalisgaon, India, created a social media storm rife with political spin. Why would villagers vandalize solar panels? One right-wing supporter commented on Twitter that the incident was intended to sabotage the prime minister's clean energy program. A left-wing party supporter reported that a right-wing MP, Ashok Saxena, was telling citizens that the villagers feared that the panels were "angering the sun-gods." Another video showed the footage and reported that the solar panels had been donated by the World Bank but that the villagers, acting under the direction of a Hindu priest, destroyed them because they competed with their gods. These ridiculous claims obscured what had actually transpired: not only were the panels in no way donated by the World Bank but the privately owned power plant had already raised controversy because it was claiming agricultural farmland. Most importantly, the video footage showed workers of the power plant protesting the fact that they *had not been paid their wages.* The narrative that invoked sun gods was a complete hoax intended to frame the laborers' actions as irrational and primitive in the face of the presumably rational and progressive endeavor to solarize the country.

supplies and competing with small- and medium-sized farmers who rely on groundwater from the same sources for their livelihoods. Additionally, extracting value from large-scale, infrastructural transformation to solar power will likely require a variety of forms of compensated and uncompensated labor, in industrial and social factories, under varying (but largely exploitative, racialized, and gendered) conditions, the world over. Finally, the project of financing, building, and operating these infrastructures will deliver considerable economic and political power to those positioned to harvest the material wealth of extractive solarity. These people rarely live with or near the externalities produced by their activity.

So much for universal, immediate access to unlimited energy that skips all the steps normally involved in producing fuel and turning it into energy. Large-scale mineral extraction and industrial manufacturing. Chemical toxicity. Energy intensiveness. Resource plunder and depletion. Territorializing occupation of lands and waters. Gender inequality. Labor exploitation. Unequal concentrations of wealth and power. At the level of its materials and infrastructure, all of a sudden, solarity sounds a lot like every form of environmental injustice that has preceded it. This is not the whole, or the only, story of solarity, but it does recommend against uncritical investment in techno-utopian imaginings in which, as Joanna Zylinska so aptly puts it in critiquing masculinist narratives of environmental tragedy and heroism, "men repair the world for me."[5]

The possibility of other solarities demands that we attend to our relationships to materials and to the infrastructures that mediate these relationships. Solarity must be about turning to face these as much or more than it is about turning to face the sun. In so doing, we are likely also to encounter alternative materialities and fugitive infrastructures that hold the possibility of plural and heterogeneous solarities and the plural and heterogeneous subjects these materialities and infrastructures mediate.[6] It is here that the truth of solarity will be found: like all mediated conditions, solarity names a condition and terrain of contingency and political struggle. Infrastructure matters, not least because, along with mediating our relationships to matter, its contingent configurations bear strongly on the sorts of subjects we might become.[7]

5. Joanna Zylinska, *The End of Man: A Feminist Counterapocalypse* (Minneapolis: University of Minnesota Press, 2018).

6. Deborah Cowen, "Infrastructures of Empire and Resistance," *Verso* (blog), January 25, 2017.

7. Nikhil Anand, Akhil Gupta, and Hannah Appel, eds., *The Promise of Infrastructure* (Durham, N.C.: Duke University Press, 2018); Antina Von Schnitzler, *Democracy's Infrastructure: Techno-politics and Citizenship after Apartheid* (Princeton, N.J.: Princeton University Press, 2018); Keller Easterling, *Extrastatecraft: The Power of Infrastructure Space* (London: Verso, 2014); Brian Larkin, "The Politics and Poetics of Infrastructure," *Annual Review of Anthropology* 42 (2013): 327–43.

Solarity as Solidarity

SOLAR MATERIALISMS remind us that *what matters* is not just our fuel sources but the relations we build and maintain with and through them. Could these be relations of solidarity? Will they be? This will be the story of solarity, just as it has been the story of petroculture, the story of all materials. Better yet: *stories,* plural. There is not one story of our relations with and through materials but many. In her account of Inuit relations to minerals in the Arctic, Emilie Cameron recounts how copper was "storied" in relation to other creatures: "By storying the copper as deer, the Dene made connections between the piece of metal in their hands and a diverse network of relations that enabled them to hunt, eat and imagine their world. Their co-existence with copper enrolled a particular network of things."[1] Similarly, Métis thinker Zoe Todd writes of nonweaponized kinship relations with carbon and fossil beings, such as oil. Todd writes, "It is not this material drawn from deep in the earth that is violent. It is the machinations of human political-ideological entanglements that deem it appropriate to carry this oil through pipelines running along vital waterways, that make this oily progeny a weapon against fish, humans, water and more-

1. Emilie Cameron, "Copper Stories," in *Far off Metal River: Inuit Lands, Settler Stories, and the Making of the Contemporary Arctic,* 84–110 (Vancouver: University of British Columbia Press, 2015).

than-human worlds."[2] The stories of oil are more complicated than critiques that universalize petroculture might lead us to believe, just as the stories of solarity are not reducible to the sun's potential as a fuel source.

To exist relationally in new ways is to take up new understandings of ourselves and our worlds, but in so doing, we will also build new knowledges experientially and observationally, experimenting with new ways of being and doing. Also, we must reorient whose knowledges we value, what knowledges we value, and what we value more generally. We must enable the retrieval of those knowledges that have been systematically laid to waste by hetero-normative-patriarchal-petro-colonial capitalism; we must aid in the recuperation of the ways of living, being, and knowing that have been violently lost through the genocidal practices of extractivism. And, when questions remain unanswered, we must collectively build new knowledges adequate to the challenges we face on this changing planet. This is to solarize: to positively disrupt and erupt dominant (hierarchical, exclusionary, and oppressive) systems and understand the world anew, in solidarity with the many and diverse beings with whom we share it.

The possibilities of solarity are sensed in real, material, and embodied ways beyond the primacy of the hypothetical and visual that industrial capitalists often ascribe to energy concerns. Bodies and beings are a key part of the infrastructures of solarity, as important as the inanimate materials set in motion within energy assemblages. Here we might consider coral reef restoration as a solar energy concern in which solarity feels like connective, embodied relations. Connectivity, and therefore solarity, can be sensed by tropical coral beings as they grow in shallow waters, their algal symbionts thriving in the filtered sunlight, providing the energy needed to exude coral exoskeletons that accrete around human-made metal frames in ex-

2. Zoe Todd, "Fish, Kin and Hope: Tending to Water Violations in Amiskwaciwâskahikan and Treaty Six Territory," *Afterall: A Journal of Art, Context, and Enquiry* 43 (2017): 107.

HOW SOLARITY FEELS

Jenni Matchett

how solarity feels | E X P A N S I V E | ungentrified | NIHILISTIC | utopian | unplanned | skeptical | speculative | like failure | illusive | vivid | disconnected | hyperconnected | intuitive | like constructive rebellion | nomadic | like a parallel universe | subtle | like the space between the time of reality and the time of ephemerality | like anarchy | like being in space, above time | like resonance | translucent | ethereal | like the experience of new beginning | poetic | instinctual | expressive | like truth | as a blank map behaves | ubiquitous | conscious | fluid | like the act of creating | like New Babylon | like the opposite of the SPECTACLE | the consideration/implication of total planetary ecological system boundaries | like grief | poetic | like an assemblage | like foraging | like pure multiplicity | variable | in flux | textured | internal | in tune | exposed | free | individual | collective | spiritual | smooth | perpetual | iterative | like a continual state of becoming

---

panding hectares of engineered reef restoration areas. The growth of coral bodies is a manifestation of multiple connections across scale, linking the global to the local, the animate to the inanimate, as well as the accumulated capital of commercial enterprises to the funding streams of university research and the economic aspirations of coastal communities. The embodied connections of solarity are thus intimately tied to questions of solidarity.

Solidarity is a complicated notion, and equity, as a possibility of solarity, is not a given in our neoliberal and neocolonial world. Just as coral beings are suffocating and dissolving in ever-warming and -acidifying seas, an increasingly catastrophic embodied condition of the production cycles of global capitalism, so, too, are coastal communities disassociating as their social and ecological systems succumb to the pressures of unprecedented change. We must not ignore the fact that there are toxic forms of solarity that spread death and destruction through connected assemblages.

The conditions that support equitable, sustainable, material forms of solarity as solidarity are, at first glance, difficult to distinguish from the toxic formations. In fact, we must seriously consider that in some instances, the forms of solarity that bring life and death are inherently and ironically entangled. Can coral survive the next fifty years of global warming without complex collaborations between the engines of capital, science, and local, communal life? Can the communities that depend on coral for their well-being do anything but adapt to the loss or reengineering of their socioecological worlds?

It is crucial to understand that the possibilities of solarity are not inevitable, even as they are complex and entangled. Solarities can be directed and shaped. Solarity is contingent and malleable. This is the promise and the pitfall of solarity and its innumerable connections. The key is to interpret and explore these solar connections in ways that enable the equitable, sustainable, and solidifying forms of solarity we want to perpetuate into the future. As reef builders—both coral and human—know, harnessing the energy of the sun requires careful attunement to shifting conditions and the coordination of many moving parts across vast scales. On the human end of the continuum, this coordination can be conducted by those who would manipulate coral assemblages for profitable exploitation or, more recently, for corporate public relations that attempt to mask continued unmitigated carbon pollution and marine resource extraction. But these schemes cannot hide the embodied evidence of the many local instances of global coral death. We must minimize the possibilities for toxicity and maximize the possibility of solidarity within embodied solar assemblages. This deceptively simple observation is as crucial for the future of coral as it is for all our planet's energy systems.

Solarity is so much more than energy—or at least, energy imagined narrowly as power, as electricity, as fuel for consumption by (some) humans. Solarity invites a much more capacious understanding of energy. After all, Earth beings are solarian. Life on this planet would not be possible without the sun, excluding some of our bacterial cousins who prefer to get their energy elsewhere. Solarity considered this way displaces the human as the primary beneficiary or consumer of the sun, drawing attention to the wondrous relationships that surround us and sustain us. As Natasha Myers puts it, it draws us to "the photosynthetic ones—those green beings we have come to know as cyanobacterial, algae, and plants."[3] The sun's generosity is not for us alone, and indeed, it is indifferent to whom it touches. But these green beings, those "sun worshippers and worldly conjurers," reveal another mode of engaging the sun—through nothing short of alchemy, they transform the world into a home for the rest of us.

Perhaps, then, it is time to learn from these plants and other photosynthesizing beings, to listen to the other beings with whom we share this world, for these are possibly our most generous cousins. Potawatomi botanist Robin Wall Kimmerer writes that "plants tell their stories not by what they say, but by what they do."[4] We are only just beginning to uncover the many magics of all that plants do, at least in Western science, even as it seems like many of their secrets will be lost forever. Plants are world creators in all senses and across scales. They weave an endless symphony from sunlight, water, and air, alchemically communing with one another and with insects, birds, and other animals that live with and among them. As Eduardo Kohn offers, forests think and act all the time.[5] They endlessly ask and answer the question of how to relate to both the sun and Earth in all their complexity, how to nurture and grow without depleting the source from which they draw, how to multiply the gift of life for beings other than themselves. So, we must choose which path to follow, which teachers from whom to learn.

Solarity invites us to learn from other beings who commune with the sun, to unearth those teachings from those who have learned to watch the trees being in the world, and who model their own

3. Natasha Myers, "Photosynthesis," *Fieldsights*, January 21, 2016, https://culanth.org/fieldsights/photosynthesis.

4. Robin Wall Kimmerer, *Braiding Sweetgrass: Indigenous Wisdom, Scientific Knowledge and the Teachings of Plants* (Minneapolis, Minn.: Milkweed, 2014), 128.

5. Kohn, *How Forests Think*.

behaviors and moral codes after these magnificent ancient beings. It invites us to learn from forest dwellers, who have been listening to and learning from the trees, and to follow their leadership in living more generously with the other beings with whom we share this world. In short, it invites humility and compassion, the values of solidarity. It demands that we take seriously that energy is not merely fuel for endless human consumption and growth but a gift that makes life possible.

Oppressive Solarities

THE NOTION OF BLEAK AND BRIGHT SOLARITIES invites us to think through the multiple affects and imaginaries associated with brightness. Orienting ourselves toward "bright spots" in the world: practices, ideas, people, institutions that might be borrowed; hacked; fostered; built upon; brought into communities, solidarities, networks. We can be receptors of these bright spots, cultivating seeds of possibility wherever they are to be found. This orientation recalls both Walter Benjamin's solar simile in "Theses on the Philosophy of History"—"As flowers turn toward the sun, by dint of a secret heliotropism the past strives to turn toward that sun rising in the sky of history"[1]—and J. K. Gibson-Graham's call to attend to actually-existing alternatives to capitalism extant in the present.[2]

Yet "brightness" is not an uncomplicated good. For example, Ousmane Sembene's 1971 film *Emitaï* features powerful scenes of old men, women, and children in French colonial West Africa who resist the demands of occupying soldiers during World War II to turn over baskets of rice or young men for conscripted labor. They are forced to sit in the sun without recourse to shade: quite literally punished by the sun. Similarly, in Ghassan Kanafani's 1962

1. Walter Benjamin, *Illuminations: Essays and Reflections,* trans. Harry Zohn (New York: Schocken Books, 1999), 255.

2. J. K. Gibson-Graham, *A Postcapitalist Politics* (Minneapolis: University of Minnesota Press, 2006).

novella *Men in the Sun,* "illegal" Palestinian migrants suffocate from extreme heat within the bowels of the water tank in which they are smuggled as they seek to travel from Iraq to labor in the Kuwaiti oil fields. Their tragedy is reinforced by the very presence of petro-infrastructures installed as the predominant energy regime amid the powerful—yet damned, hellish, and deathly—heat of a sun-"rich" region. Such an infernal scenario intensifying is powerfully rendered in Kim Stanley Robinson's speculative novel *The Ministry for the Future* (2020), in which the citizens of South Asia have to exist in a near-future world under brutal and yet commonly excessive wet bulb temperatures. These are powerful reminders that the sun is not only a precondition for earthly life but can also be oppressive, overwhelming, and hostile to life. In some planetary contexts, where the sun's reach is strongest, brightness and bleakness are intertwined. Excessive heat is not merely unsupportive to life (particularly human and mammalian life) but actively kills. In the context of our ever-warming planet, large swathes of Earth are becoming uninhabitable for most humans and nonhumans, especially in the tropical belt that is home to a large part of the human population and much of the biodiversity of Earth.

A few statistics are instructive in illustrating the scale of the problem: approximately 3.3 billion people, or 40 percent of the human population, live in the tropical zone. This includes 85 percent of the world's poorest people, who are also among those most susceptible to the effects of climate change while contributing least to its causes. Although human bodies are efficient at regulating internal temperatures, this becomes harder as temperatures and humidity rise. From the perspective of human health, something called the "wet bulb temperature" is critical. *Wet bulb* refers to the temperature a thermometer covered in a wet cloth would measure. It matters because it reflects the maximum amount of cooling that can be achieved by evaporation. It is critical because wet bulb temperatures of higher than thirty-five degrees Celsius are life threatening to humans: unable to cool down, our bodies would succumb to heat stress. This is not merely a speculative account of an uninhabitable distant future; since the 1980s, there has been a fiftyfold increase in the number of places experiencing extreme heat. In tropical zones, where temperatures and humidity are both high, the number of people experiencing heat stress has been growing every year, especially among those workers who are involved in outdoor labor, such as in construction and agriculture.

In these cases, the inequalities wrought by neoliberalism are exacerbated manifold by global warming. For instance, agricultural workers in the United States, the majority of whom are migrant workers from Mexico and other South American nations, are forced to work in conditions of extreme heat.[3] In 2020, at least twenty-one days were too hot for safe outdoor work. It is estimated that by 2050, farmworkers will meet unsafe daytime summer temperatures on at least thirty-nine days of the harvest season. This is the case around the world, where extreme heat and other weather conditions caused directly or indirectly by rising temperatures result in extreme floods and extreme droughts, both of which are further linked to different health conditions. For instance, Alex Nading writes about the epidemic of chronic kidney disease among sugarcane workers in northwest Nicaragua and other Central American nations.[4] While work in the sugarcane plantations has always been deadly, Nading writes, chronic kidney disease is a recent phenomenon. Studies suggest a strong correlation between the disease and heat exposure, indicating a kind of "thermal tipping point" that cane workers are now experiencing. While cane (and other colonial) plantations have always been suffused with heat as well as violence, the rise in chronic kidney disease indicates an intensification of these combined violences, in which increased heat alongside extractive work

3. Tim Radford, "US Farm Workers Face Worsening Lethal Heat," *Climate News Network,* May 6, 2020, https://climatenewsnetwork.net/us-farm-workers-face-worsening-lethal-heat.

4. Alex Nading, "Heat," *Fieldsights,* April 6, 2016, https://culanth.org/fieldsights/heat.

conditions (which include fewer breaks for shade, water, and rest) results in an embodied experience of slow violence.

The experience of oppressive solar heat is not limited to construction laborers and farmworkers. Jason de Léon writes, in unsparing detail, of the ways in which the U.S. government has weaponized the extreme heat of the Sonora Desert as a "natural" defense and deterrent against migration.[5] The Border Patrol officers deliberately close off the safer points of access to the United States, directing people seeking a safer life toward the harsh, burning landscape of the desert. This policy of so-called prevention through deterrence is in fact a necropolitical strategy of weaponized solarity. As de Léon writes, the heat in the desert not only kills the undocumented migrants seeking a better life in the United States but also destroys evidence of their lives, bleaching, wearing away, and breaking apart their bodies and possessions. De Léon exhorts readers to look at the "cruel, brutal affair" of desert border crossing that dehumanizes and murders migrants, to witness and acknowledge the slow and painful violence of hyperthermia, dehydration, heatstroke, and other ailments that they are forced to endure. This is a reality of solarity that we, too, cannot shy away from. While we often celebrate the sun's generosity and abundance, these stories are important reminders of the deep inequalities in receiving this abundance, especially in cases where there is no respite. Abundance is not always an untroubled good.

There's something important and appealing about the nuance and structuring ambivalence that a stance at once toward yet away from the sun demands. One needs gray spots, clouds, shade, relief from the sun, as much as "bright spots." "Blue-skying" offers an apt metaphor for dreaming big, grasping after the possible without constraint or limit, but it also represents the possibility of a nightmare: a sun without limit. Here there is something powerfully

5. Jason De León, *The Land of Open Graves: Living and Dying on the Migrant Trail,* California Series in Public Anthropology 36 (Berkeley: University of California Press, 2015).

suggestive in considering the technics of how solar panels actually work. Sunlight excites electrons, setting them in motion, bouncing around. Yet some 80 percent of that movement is nonproductive; only around 20 percent gets channeled into usable electricity. Rather than prompting dismay at such "inefficiency," this might be a powerful gesture toward something like degrowth, against demands for power and efficiency, against the demand that we be productive all the time. Production is not the same as creativity. Some people need less sun, not more. Could it be that protection from excess and contentment with untapped potential energy point the way toward an anti-manifesto of solarity? Perhaps, in embracing "intermittency," we can find solace, rest, regeneration, creativity, a conscious turning away from the exposed regime of excessive heat and its forbearances.

Decolonial and Feminist Solarities

INDIGENOUS PEOPLES are among those who are most adversely affected by fossil fuel extraction practices, such as fracking, that poison nearby waters, land, and living beings. This form of exploitative colonization will continue for as long as fossil fuels are being burned. Oil is not sustainable, neither as a finite resource nor as an industry, a product, a cultural institution, or a social obsession. Fossil fuels are so relied upon that it is difficult to imagine a future world that is not made possible by carbon-made energies. Rethinking energy usage through a newly formulated relationality with the sun is a needed step toward building better, more sustainable futures for Earth. Solarity might provide answers to only one aspect of how energy production and consumption are reformulated. However, it is a partial solution to the multifaceted problems of petroculture and global capitalism. Indigenous voices must be centered in discussions about the futures of territories with which they have lived in relation since time immemorial, including determining how the lands and waters are being used for energy production. Decolonization must be a large part of the speculative scope of futures after oil.

The promise of solar to address the problem of climate change and to mediate fundamentally transformed social and political relationships and organizational forms is most evident in situated practices of solarity on the ground in diverse locations. Potawatomi scholar and activist Kyle Whyte observes that Indigenous people "confront climate change having already passed through environmental and climate crises arising from the impacts of colonialism."[1] Whyte quotes Heather Davis and Métis scholar Zoe Todd, who write that the environmental crisis some have named the Anthropocene is "really the arrival of the reverberations of that seismic shockwave into the nations who introduced colonial, capitalist processes across the globe in the first half-millennium in the first place."[2] It is therefore not surprising that some Indigenous communities are leading the way in terms of developing sustainable, renewable energy infrastructures on their territories, as part of their efforts to transition away from costly, unreliable, emission-intensive, diesel-generated electricity.[3]

Crucially, these initiatives are not just about saving money and saving the planet. They are also intrinsic to an assertion of material sufficiency and political self-determination by these communities, in the context of ongoing struggles for decolonization and in light of histories in which energy and other infrastructures have been deployed as instruments of dispossession and deprivation by settler-colonial states and industries.[4] Melina Laboucan-Massimo, a member of the Lubicon Cree First Nation and a scholar-activist who led the Piitapan Solar Project in Little Buffalo, Alberta, puts it directly: "We are tired of being economic hostages in our own homeland. We're not looking for a clean energy grid that's owned by big corporations like Suncor or Enbridge but by communities that

1. Kyle P. Whyte, "Indigenous Science (Fiction) for the Anthropocene: Ancestral Dystopias and Fantasies of Climate Change Crises," *Environment and Planning E* 1, no. 1–2 (2018): 226.

2. Heather Davis and Zoe Todd, "On the Importance of a Date; or, Decolonizing the Anthropocene," *ACME: An International Journal for Critical Geographies* 16, no. 4 (2016): 774.

3. Emily Gilpin, "Skidegate on the Way to Becoming a 'City of the Future,'" *National Observer,* April 9, 2018, https://www.nationalobserver.com/2018/04/09/brighter-news-clean-energy-success-story.

4. Maryam Rezaei and Hadi Dowlatabadi, "Off-Grid: Community Energy and the Pursuit of Self-Sufficiency in British Columbia's Remote and First Nations Communities," *Local Environment* 21, no. 7 (2016): 789–807.

PIITAPAN SOLAR PROJECT IN LITTLE BUFFALO, ALBERTA

In 2015, Sacred Earth Solar (formerly Lubicon Solar) launched the Piitapan Solar Project, a 20.8-kilowatt renewable energy installation in Little Buffalo that powers the community health center.* The eighty-panel solar project has created more green jobs and reduced the community's reliance on fossil fuels. The Lubicon Lake Band is a Cree-speaking community whose reserve lands are centered on the village of Little Buffalo, Alberta. This northern community, about a five-hour drive from Edmonton, is the home of Melina Laboucan-Massimo, the founder of Sacred Earth Solar. Sacred Earth Solar not only helps install solar panels in Indigenous communities of Turtle Island but also trains members of the communities in installation and maintenance. Members of the community describe the pride and sense of self-reliance they experience from having the solar project in their town. Located in Alberta, home to the ecologically devastating oil sands, the success of this project is a testament to the possibilities of solar for bringing about social, ecological, and energy justice even in the petrocultural capital of Canada.

* https://sacredearth.solar/piitapan.

actually own their power."[5] Many of these projects entail sophisticated modes of governance, political organization, and relationships with land, place, and nonhuman others that far exceed anything of which Western, liberal democracy has so far been capable.[6] Clearly "we" have much to learn from those who are actively Indigenizing solarity.

We could say much the same of other categories of people historically marginalized by regimes of carbon energy and their enabling infrastructures and who are now developing solarities that hold open the possibility of more responsible and egalitarian forms of social organization. For example, Shane Brennan tells the story of

5. Quoted in Richard Thompson, "New Economy Trailblazer: Melina Laboucan-Massimo," *Rabble.ca*, November 7, 2017.

6. Vanessa Watts, "Indigenous Place-Thought and Agency amongst Humans and Non-humans (First Woman and Sky Woman Go on a European World Tour!)," *DIES: Decolonization, Indigeneity, Education, and Society* 2, no. 1 (2013): 20–34.

Soulardarity, a community organization led by African American residents in Highland Park, Michigan, where, in 2011, the local utility company removed the town's streetlights in response to declining revenues. In response, Soulardarity activists worked to "install solar-powered lights that are collectively paid for, owned, and controlled by local residents in the model of a sustainable lighting cooperative."[7] Brennan describes the result as an example of "visionary infrastructure . . . a form of material and social practice in which the collaborative work of building critical infrastructures is inseparable from the imaginary work of collectively envisioning the future with and through those infrastructures."[8] Visionary infrastructures like the Highland Park solar streetlights convene communities, and "once a community has been convened, a visionary infrastructure then helps this community imagine and create alternatives to the existing system."[9]

Similarly, Dagmar Lorenz-Mayer reminds us that "feminist solar imaginaries" have been in circulation since at least the 1970s, when feminists working at the intersection of antinuclear, peace, and environmental activism "articulated ideas of socio-environmental justice through the envisioning of commons-oriented practices where systems of 'renewable energy . . . belong to the people and their communities, not to the giant corporations which invariably turn knowledge into weaponry.'"[10] This does not mean that solar is a panacea for women—Lorenz-Mayer's study of the exclusion of Roma women workers at a Czech PV installation indicates otherwise—but it does point to the possibility of, and urgent need for, alternative solarities informed by "feminist imaginations of community, participation and care" that generate "an ethos of

7. Shane Brennan, "Visionary Infrastructure: Community Solar Streetlights in Highland Park," *Journal of Visual Culture* 16, no. 2 (2017): 168.

8. Brennan, 176.

9. Brennan, 178.

10. Dagmar Lorenz-Meyer, "Becoming Responsible with Solar Power? Extending Feminist Imaginings of Community, Participation and Care," *Australian Feminist Studies* 32, no. 94 (2017): 431.

querying and imagining community-run solar power, as well as a responsive non-interventionist attentiveness to forms of life, incorporating yet indifferent to humans, that expands the contours and possibility of an ethos of care and communing."[11]

Such projects and imaginings might be marginal to the extractive, racist, and masculinist solarities that global capitalism has in store for most us, but they remain central to the possibility that things could go otherwise. Sketching a "feminist counterapocalypse" that unsettles finalist thinking and its moralizing, depoliticizing tendencies, Joanna Zylinska seeks to "engender a more anchored, embodied, and localized sense of response to, and responsibility for, the milieu we earthlings call home. 'The end of man' pronounced as part of the current apocalyptic discourse can therefore be seen as both a promise and an ethical opening rather than solely as an existential threat."[12] Stepping into this opening involves aesthetic practices that explore "better ways of sensing the Anthropocene . . . to produce a more engaging and meaningful encounter beyond the shock and awe effect of the postindustrial sublime." These aesthetics—potentially the aesthetics of a feminist solarity—"embrace precarity as a political horizon against which the dream of infinite linear progress is presented as expired," while refusing to relinquish the "drive for justice" that will be required to make solarity livable.[13]

There is a reason that some of the most promising glimpses of solarities today come from Indigenous communities who continue to most acutely experience the unevenly distributed consequences of a fossil economy propelled in many regions by settler-colonial infrastructures. Solarity offers a potentiality of decolonial justice against the stratification inherent in the production and consumption of fossil fuels. Projects like Laboucan-Massimo's Piitapan Solar Project do more than address energy deficits at the level of tech-

11. Lorenz-Meyer, 440.

12. Joanna Zylinska, *The End of Man: A Feminist Counterapocalypse* (Minneapolis: University of Minnesota Press, 2018), 7.

13. Zylinska, 65.

nology—a process that risks reproducing the same stratification found in the fossil economy. Instead, these projects are a practice of worldbuilding and futurecasting in resistance to the present energy regime. The fossil economy was imposed from above by a dominant class who recognized in fossil fuels the potentiality to deepen inequities and concentrate wealth. Solarity seems to hold the promise of upending such concentrations. If solarities are to be just rather than unjust, they must be generated from below rather than from above; solar energy must be as dispersed as the sun's rays, refusing the kinds of concentration that petrocapitalism has engendered.

Light and heat come freely to Earth. They are the basis for planetary life. The conversion of the sun's rays into chemical energy is the initial food source, the basis for the trophic cascade that weaves together complex chains of abundance and interconnection that characterize healthy ecosystems. Left to itself, the sun models an economy based on abundance, on gifting, on interconnection, on multispecies flourishing. This is an economy of cycles, diurnal and seasonal. It is a dynamic economy of constant circulation, constant redistribution of life force and energy: upwellings of ocean currents, the jet stream flowing thousands of miles, birds on their winged migrations connecting the Arctic with Africa. Yes, there is scarcity (hunger when the caribou do not come; parched soils in seasons without rain). Yes, there is competition (light-seeking saplings below the forest canopy; algal blooms blanketing waterways, obscuring access to oxygen and sky). But at its core, the solar economy is one of abundance and renewal, of plenty.

Indigenous lifeways of Turtle Island and beyond acknowledge and honor the gifting that generates this abundance, whether through the Thanksgiving Address of Haudenosaunee peoples—which expresses gratitude for and acknowledges our interdependence with the beings and energies that make up Earth—or through the potlatch system by which Northwest Coast peoples redistribute wealth, regenerating social relations. Indigenous economies, taking their cue from creation stories where the energy of life is given freely, share abundance to produce more abundance. The bounty

FEMINIST SOLARITIES: SPECULATION, INTERRUPTION, INFRASTRUCTURE, SOLIDARITY

Jessie Beier

Reimagining energy systems around the energy of the sun calls on thinkers, makers, and feelers to imagine and practice solarities that are intentionally and explicitly feminist and, perhaps more important, to demonstrate how feminist solarities can offer the much-needed tools, both conceptual and material, to interrupt the given, reengineer infrastructures, speculate on the possible, and, ultimately, reimagine collectivities characterized by unthought sol(id)arities.

Feminist solarities suggest four main lines of inquiry: speculation, interruption, infrastructure, and solidarity. Drawing from both historical and emerging theories and practices of feminist thought, we seek to investigate the relations between feminism and energy transition as they are situated within today's particular ecological and social crises.

Our speculations led to the creation of a collective conceptual persona called the Mirrorland Collective, a name drawn from Dagmar Lorenz-Meyer's explorations of feminist technoecological dis/articulations and her call to question what it means to become "response-able" in relations with solar power.* Prompted by this question, the newly formed Mirrorland Collective authored and sonically recorded a manifesto titled "A Big Pile of Glitch: A Manifesto for Feminist Solarity."†

Our experiments taught us that reorienting energy in relation to feminist sol(id)arities involves developing speculative modes of working and thinking together, modes that necessitate unthought recalibrations and coordinations to one another, to nonhuman forces and intensities, to the cosmos, and, specifically, to the sun, that stellar ball of lightning burning in the sky.

* Lorenz-Meyer, "Becoming Responsible."

† https://www.justpowers.ca/app/uploads/FeministSolarities Manifesto_FINALv2-1.pdf.

of Earth, which is and must be rebirthed and regenerated over and over, is stewarded and supported using principles of interconnection, reciprocity, and balance.

In practice, this means initiatives like Indigenous Clean Energy's 20/20 Catalysts Program, a hands-on project that, through in-class learning, mentoring and coaching, site visits with energy experts, and webinars and online support, helps nurture capacity in

Indigenous communities embarking on clean energy projects. Solar energy becomes a site not just for cultivating energy independence—an important goal in communities where the costs of diesel (to health as well as finances) can be crippling—but for supporting Indigenous governance and cultural renewal. Solarity offers the possibility of energies that are less extractivist, that can circulate through noncapitalist relations, and that can fundamentally challenge the principles of scarcity and competition foundational to capitalist economics. In this way, solarity seems to posit a fundamental challenge to capitalism.

This is a possibility, not a given. As we noted earlier, our energy mix could shift toward a solarity that does not disrupt the status quo: large solar farms, linked to the existing energy grid, owned by major corporations that have gained near-monopoly status through employing economies of scale; PV manufacturing that turns a blind eye to unethical sourcing of rare earth minerals from areas where working conditions are dangerous, wages are appallingly low, and competition for control of mining fuels armed conflict; a "rentier" economic model, in which the surplus generated by abundance is not distributed to cultivate widespread, interconnected flourishing but channeled to further the accumulation of a small elite, who grow their power by charging fees for access to the resources they have captured. Solarity is rife with such contradictory possibilities.

As renewable energy tech shows increasing promise of fueling a life after oil, envisioning a future beyond extraction remains a difficult task. Centuries of extractive logics have conditioned an extractivist line of sight that sees in what we might call "progress" (or even simply a "good life") a necessarily extractive foundation. Extraction, in other words, appears as a nonnegotiable precondition to what comprises many visions of a good life, postcarbon or not. Current PV infrastructure, for instance, requires lithium and other rare earth minerals in its construction. And wind turbines are primarily constructed from steel, a material whose extractive legacy played a crucial role in cementing the industrial age. Embedded in the materials many hope will fuel transition, extractivism as an impulse and, indeed, ideology premised on nonreciprocal relations with the human and nonhuman world continues to haunt visions of postextractive futures. Yet, a core dimension of the promise of solarity rests precisely on its activation of ways of being and seeing beyond the extractive, a vision from scarcity to abundance that reckons with extractivism in the first instance. Conventional wisdom tells us that we shouldn't let perfect be the enemy of good. In a future conditioned primarily by principles of solarity, particular extractive relations may remain as material residues from our extractivist past and present, but so, too, will emerge new and previously subdued possibilities for relations of reciprocity.

Yet, even all "small solarities" might not be inherently "good" or wholly "bright," and here it is important to resist the impulse to romanticize instances of small-scale, community solar energy as being somehow pure and untainted. Whatever use solar panels are put to, it is important to trace the means by which they are produced and the relations of extraction, pollution, and dispossession that are required in the course of their production. The extraction of lithium, rare earth metals, and other materials required for the creation of solar panels often takes place in precarious and hazardous laboring conditions that necessitate large-scale displacement of Indigenous and poor people. In this case, the self-determination of one small-scale solar community might be intertwined with the continued dispossession and oppression of another. In this case, we need to ask, is there a possibility of ethical extraction and mining? Utopian, anticapitalist solarities would require the creation of new, ethical processes of mining and resource extraction.

Brooklyn-based environmental justice organization Uprose is poised to become New York State's first solar co-op. Working with multiple partners and cutting through the city's red tape have been challenging, but Uprose has persisted because the establishment of a solar cooperative would not only bring concrete economic benefits to the predominantly working-class, immigrant community in Sunset Park but also be a tangible example of community self-determination and empowerment. As Uprose executive director and longtime climate justice activist Elizabeth Yeampierre put it, "the idea for community-owned solar comes out of efforts from the climate justice community to operationalize just transitions. To basically make it possible for our communities to start moving off the grid and to start creating utilities or mechanisms that will help them thrive in the face of climate change."[14] Sunset Park Solar is a momentous victory for renewable energy: fewer than 1 percent of New Yorkers' homes are powered by solar, and modern renewables provide less than 5 percent of power in New York State, despite Governor Cuomo's much-ballyhooed initiative to "Reform the Energy Vision."[15]

Yet, while seeking to breathe life into lofty ideas about just transition, Uprose's community solar project is unfolding in a predominantly Latinx community that knows very well the links between resource extraction and contemporary empire. As Yeampierre argues, the solar co-op project is about more than just knowing how to install solar, or even understanding just transition. It is, she states, about "moving away from extraction to a different kind of life."[16] Yeampierre, who is originally from Puerto Rico, is well aware of the links between energy and colonialism, particularly following the systematic attack on public power that took place before and after Hurricane Maria devastated the island. For activists like Yeampierre, just transition cannot involve simply switching from fossil fuels to renewables while maintaining the social hierarchies constructed by fossil capitalism. Just transition also involves challenging the current extractive economy, which reaches far beyond fossil fuels alone. If solarity is to be sustainable, it must also be decolonial and anticapitalist.

14. Lourdes Pérez-Medina and Elizabeth Yeampierre, "The People's Power," *Urban Omnibus,* April 10, 2019, https://urbanomnibus.net/2019/04/the-peoples-power/.

15. "Reforming the Energy Vision (REV)," New York State Government, March 30, 2015, https://www.nypa.gov/innovation/initiatives/rev.

16. Pérez-Medina and Yeampierre, "People's Power."

Solar Temporalities

IT TAKES ABOUT EIGHT MINUTES for the energy of the sun to reach Earth. That light and heat have history, speed, and trajectory confirms that solarity is a temporal condition. In the European experience, the arc of solarity's past, present, and future is a familiar one: the past is the era of fossil fuels; the present is a time of urgency and transition; and the future is the time of deliverance and rebirth. This makes it easy to fall into the timeline, but what if the temporalities of solarity actually step out of line? Here we are reminded that solar energy is the primary, fundamental source of energy before fossil fuels both historically and metabolically. Without prior photosynthesis, there would be no carbon to combust. Solarity is both the "before" and the "after" of fossil fuels.

The reality of climate change confronts us with the fact that the era of fossil capitalism, in which carbon dioxide accumulates in step with profits, is not now and may never be behind us, never simply buried. It is more like a zombie that refuses to die or a ghost that will forever haunt us.

As a source of "clean" renewable energy, solar articulates easily with the urgency attached to the present climate crisis. Climate change generates a global, nonnegotiable imperative to hasten development and implementation of near-zero-emission energy infrastructures at an industrial scale. Environmentalists have a name for this imperative: we call it "energy transition," and we, too, think it cannot wait. It is not surprising that environmental advocacy around energy transition, including to solar, is animated by a sense of urgency.

However, urgency also serves strategies for energy transition that entrench, intensify, and depoliticize environmental and other injustices. The environmental crisis capitalists who stand poised to profit from solarity are fluent in the language of urgency. Indigenous, colonized, racialized, and impoverished communities have lived, for decades and centuries, with ongoing environmental emergencies produced by the economies and ways of life whose survival is now presented to them as a global imperative. These are structural conditions that have never commanded global attention in the way climate change does, and the particularity of the experiences arising from them is erased by the totalizing urgency of the climate crisis. This is why the transition to solarity must be approached not merely as an urgent question of global energy transition but as a matter of global environmental justice that has been a long time coming.

The absence of diurnal time across the circumpolar world, for instance, asks specific questions of the sun as an energy source. If solarity is premised on a relational constitution between human practices and particular forms of energy mediation, then it follows that the largely Inuit, Inupiak, Sámi, and other Indigenous communities who have long resided in place across the Arctic experience the sun as a starkly seasonal transit, and one not diurnally tied to energy abundance. Seasonal changes, and the light conditions that accompany them, often sweep over lands, waters, and ices in accelerated rhythms, with an energy-giving sun making way for the more energy-static moon. In Iqaluit in April, dusk settles in like a slow-fading color gradient, tinting the city's buildings a purply blue then quickly flicking to black. These are places where stark distinctions between dark and light become part of the "lived ordinary." If, according to Lauren Berlant, infrastructure can be thought of as the "living mediation of what organizes life," that is, the "lifeworld of structure,"[1] then diurnal time is intimately bound up in this vital

1. Lauren Berlant, "The Commons: Infrastructures for Troubling Times," *Environment and Planning D* 34, no. 3 (2016): 393.

process of mediation and the potential interventions in characterizing southern-oriented energy infrastructures it can make. Energy deficits across the circumpolar world, and communities' reliance on diesel-powered forms of electricity to heat homes, operate under a decidedly ambiguous midnight sun—a signal that Bataille made out as an "extreme incandescence"[2] and an emblematic part of the planet's cyclical accumulation and expenditure of solar energy.

Yet what do diesel storage tanks, ubiquitous mushroom-like volumes that dot the outskirts of communities across the circumpolar world, figure under the energy-absent winter night sky? Indeed, the lived ordinary of residents is bound to an understanding of solarity that finds this incandescence in summer tundra and taiga, in the mosquitos and old, dry mosses of summer hunting camps. This, too, is energy without the possibility of infrastructural mediation. Historically, and it seems for our warming present, southern projections of the circumpolar world's solar imaginaries are such that the slowly drawn out and then swift transit between light and dark forecloses a merely infrastructural conception of solar energy. Solarity can only assume the unlimited potential of the sun as an energy source under the climate conditions of diurnal time. Being in place across the inhabited Arctic, in both light and dark, is a reminder that incandescence and energy are not always bound together as infrastructure.

Solarity's futures are similarly complex. As Rhys Williams writes, "solar is so resonant of a 'fresh start' that it slides into ideas of a 'clean slate.' Solar futures are attractive in part because they offer the chance to forget what came before, to absolve us of our own environmentally-damaging history, or at least to shield ourselves from it."[3] For many of us, solarity's promise of amnesia—the prospect of forgetting the past and proceeding as if it never happened—

2. Bataille, *Accursed Share*.

3. Rhys Williams, "'This Shining Confluence of Magic and Technology': Solarpunk, Energy Imaginaries, and the Infrastructures of Solarity," *Open Library of Humanities* 5, no. 1 (2019): 13.

INDIGENOUS SOLARITIES

Maize Longboat

My contribution to After Oil School 2: Solarity was as a co-instructor for the "Imagining Indigenous Solarities" workshop given alongside Waylon Wilson and as a representative of Aboriginal Territories in Cyberspace (AbTeC[*]) and the Initiative for Indigenous Futures (IIF[†]). Cofounded by Jason Edward Lewis and Skawennati, AbTeC is an Aboriginally determined research-creation network whose goal is to ensure Indigenous presence in the web pages, online environments, video games, and virtual worlds that compose cyberspace. AbTeC manages IIF, a partnership of universities and community organizations dedicated to developing multiple visions of Indigenous peoples tomorrow in order to better understand where we need to go today. Through its four main components—workshops, residencies, symposia, and archive—IIF encourages and enables artists, academics, youths, and elders to imagine how we and our communities will look in the future.

In the "Imagining Indigenous Solarities: 7th Generation Character Design" workshop, Waylon and I invited participants to imagine a future informed by solarity and the Indigenous future imaginary discourse. Questions like "what might our descendants be like seven generations from now?" and "what about the world they live in or how society might function?" guided our group of eleven participants through the discussion portion of the workshop on the first day. On the second day, we moved into the workshop's hands-on portion and gave participants the opportunity to imagine and design characters from their futures as a method of critically analyzing our society today. By imagining the future, we are able to better understand our individual and collective responsibilities to our descendants, thereby providing us with a base for enacting real change in the here and now.

Our group discussions began after offering a territorial acknowledgment and then posing the question of what a territorial acknowledgment actually is. Many of us agreed that these formal acknowledgments are as much an affirmation of continued Indigenous presence within a certain location as they are a method for everyone to position her own relationship to that location personally. We found that this helped us all to build up a sense of responsibility to place that we could continue to grow moving forward. Spaces meant for intellectualizing and theory crafting in academia do not encourage scholars to bring their positionalities into their research. In this case, we asked workshop participants to think about their futures and creatively speculate how their imagined descendants exist in a world after oil.

The character sketches that were created as part of this workshop were part of worlds with diverse futures. I imagined a world in which my descendant lives where humans have formalized their relationality to the sun by becoming biologically bonded to solar energy. These "solar-cyborgs" carry their own miniature suns within them and are in turn deeply tied to solar cycles—cultivating gardens in spring and summer, while resting in fall and winter. All of the contributions were unique and offered alternative perspectives on a world after oil. Some considered the sun to play a large part in the worlds of their descendants, while others focused their attention on other forms of energy production, restructuring of society, and kinship. Positioning how oil personally impacts us presently, as both Indigenous and Settler peoples, is a vital first step in realizing a world that we want for our collective descendants.

* https://abtec.org/.
† https://indigenousfutures.net/.

---

might be even more attractive than its promise of limitless energy. Interestingly, this is precisely the move to innocence by well-meaning environmentalists that many Indigenous and decolonial thinkers and activists refuse, even and especially under the urgent pressure of colonial and settler-colonial capitalism's climate emergency. It is not just that clean-slate solar (i.e., "let's make a fresh start and work together to save the planet with solar energy!") erases the experience of, and responsibility for, historical and ongoing environmental injustice committed against Indigenous and colonized populations, lands, waters, and species. It's that a future in which the past is forgotten is inimical to the temporal orientations many colonized and Indigenous peoples bring to their environmental relations. When Kyle Whyte describes "a spiralling temporality in which it makes sense to consider ourselves as living alongside future and past relatives simultaneously as we walk through life,"[4] Whyte refers to a sense of time in which the past is always already present in the future. In Whyte's account, spiraling time "may be lived through narratives of cyclicality, reversal, dream-like scenarios, simultaneity, counter-factuality, irregular rhythms, ironic un-cyclicality, slipstream, parodies of linear pragmatism, eternality,

4. Whyte, "Indigenous Science (Fiction)," 228–29.

among many others."[5] This speaks to a solarity that is something other, something more, than the fantasy of deliverance, something like the ongoing work of energy responsibility.

For those not accustomed to spiraling time, the work of energy responsibility might include recovering past ways of being and relating to the world and to each other. In a famous essay, the British historian E. P. Thompson argued that the transition to an industrial society restructured the outer habits and inner lives of common folk in eighteenth- and nineteenth-century England. Industrial capitalism produced more time-disciplined workers (though not without resistance), separated labor from leisure, and commodified time itself. Speculating in the twentieth century about an automated future of greater leisure, Thompson reflected, "Men might have to re-learn some of the arts of living lost in the industrial revolution: how to fill the interstices of their days with enriched, more leisurely, personal and social relations; how to break down once more the barriers between work and life."[6] "Values," wrote Thompson, "stand to be lost as well as gained" over the course of time.[7] Some of them are worth recovering.

Turning to the pasts of different societies and cultures could inform solarity in a variety of ways. We might "relearn" more restorative, leisurely orientations toward time; alternative ways of relating to others; what the poet Gary Snyder called "the inherent aptness of communal life";[8] the sound of wind, water, and birds before combustion engines; closer relations to the land and water; the skills and arts of attention lost to automation; natural abundance living with the sun. What might we relearn, for instance, from the agro-ecological success of early farming communities in New

5. Whyte, 228–29.

6. E. P. Thompson, "Time, Work-Discipline, and Industrial Capitalism," *Past and Present* 38, no. 38 (1967): 95.

7. Thompson, 94.

8. Gary Snyder, *Turtle Island* (New York: New Directions, 1974), 91–92.

England? Or from the notions of property of Indigenous societies in North America? Or from the legacy of biodiversity cultivated by India's women farmers?

This is not a nostalgic call to recreate the past. It is far too flawed for that, and those who invoke it in the service of violence and oppression must be resisted. Moreover, the legacies of modernity cannot simply be erased, and some of its achievements we will want to retain. In a solar future, we will be able to produce forms and amounts of energy and technologies unimaginable in the past. We will carry the memory of injustices that ought not to be forgotten. Relearning from the solarities of the past with an eye to the future will generate new forms of living and relating that are at the same time grounded and open to change.

According to the best climate predictions available, what we do—or fail to do—within the next decade will have monumental effects for the future of life on planet Earth. Yet, in facing what is perhaps the single greatest challenge our species has known, our responses as individuals and collectives are marked by temporalities of procrastination and denial that could very well prove cataclysmic on a scale not experienced in historical memory. These are most pronounced in the Global North, where visions of prosperity, freedom, and progress have been shaped for the past two centuries by the very fossil fuels from which we now must orchestrate a calculated, collective exodus. As the carbon clock edges closer to devastating and irrevocable levels of global warming, those nations that have benefited most from the age of hydrocarbons find themselves beset by reactionary tendencies that can only appear, for anyone who has apprehended the scope and urgency of the current need for energy transition, as pathological forms of distraction and disavowal. As we witness the uncanny return of such twentieth-century bugbears as the threat of nuclear war, white supremacy, xenophobia, populist nationalism, renewed assaults on women's reproductive rights, and the continued upward appropriation of all remaining forms of social wealth, we must question whether such spectacles, alarming as they are, do not actually provide performative, therapeutic distractions from the colossal, collective task of building a new, just, and sustainable world.

This tendency points toward a crucial fact of our current predicament: although we possess ample technological capabilities for enacting the necessary energy transition, we lack narrative technologies for imaging and conceptualizing such a shift. It is the humanities rather than the hard sciences that occupies a privileged position to address this gap in our cultural, political, and historical sensibility. The response to global warming can only issue from beyond the horizons that fossil fuels have rendered hazy and suffocating and will require the efforts of historians, cultural analysts, sociologists, poets, and artists as much as engineers and scientists. It will require new lexicons, narratives, and languages and a new name for the collective potential for transformation offered by the current junction. Solarity is one such name, pregnant with uncertainty, novelty, and possibility.

The Work of Solarity

RADICAL SYSTEM CHANGE, such as that imagined through solarity, requires honest and integral relationships to our emotions. Theorizing solarity requires a continued engagement with the types of emotional work and affective responses that these processes invoke. In other words, energy scholarship—on solarities or otherwise—must take seriously the emotional states in which people live, theorize, and enact environmental justice over the *longue durée*.

For many, energy transition away from oil can be threatening. It can destabilize people's sense of identity, values, and everyday ways of life by challenging the (often unexamined) socioeconomic practices and energy resources underpinning them. Imagining—let alone building toward—a post-oil future involves deep personal and collective reckonings with our attachments to the present. And these reflections require real emotional work. In most oil-producing societies, discourses about energy transition can spark feelings of optimism and excitement about a green future, joy in community collaboration, fear and anxiety about deprivation or being made redundant in a new economy, and even anger about a loss of choice or power. Energy transition therefore requires us to engage with these interconnected and sometimes contradictory emotions—and the significant forms of affective labor this work requires.

Building a post-oil future will require more than the development of new technologies to, say, generate and store immense amounts of solar power and adapt preexisting energy grids. We will have to engage with "our" affective entanglements with resources like oil, wind, or solar in the now. We will need to contend with the ways they have historically shaped our landscapes, communities, and emotional lives. A transition to energy from the sun will demand our affective energies, our emotional labor, and intellectual and physical work. We will need to shore up ourselves and our communities with care as we continue to demand systemic change and accountability from governments while struggling with the debilitating compromises we must all make when living within petrocapitalism.

To put it another way, any possibility of a future solar community is contingent on seeding and cultivating joy. The more just, sustainable relations we might envision in solarity depend as much on the sustainability of our capacity for joy as they do the renewable power of the sun.

Transition to solarity will also require mediation. It will require a set of techniques, infrastructures, architectures, technologies, texts, images, and affects that move people and places from being out of relation with the sun into sustainable relation with the sun. In some cases, this might be a process of disintermediation, a process of removing elements that might have stood between an entity and the sun or obscured one's connection to the sun. This would be the case, for example, in the creation of forms of open-air architecture that provide more solar exposure for bodies, for plants, and for growth. In other cases, a move toward solarity requires the introduction of new materials that refract or reflect solar emissions such that they become newly accessible for populations. Here the solar panel is a prime example—a medium that accepts solar emissions and then transforms and directs them, as the resource of energy, into the infrastructure of the grid. For yet others, it will require media that protect them from unwanted and punishing exposure to the sun's relentless heat.

Achieving solarity will also require politics. In "Who's Afraid of Democracy?,"[1] Geoff Mann recounts an experience of attending a meeting in Vancouver organized by the Canadian Centre for Policy Alternatives (CCPA). Involving a large number of individuals and organizations with interests in environmental and social change (including representatives from nongovernmental organizations, unions, and universities), the discussion at the event focused on the possibility of a "just transition" as part of a shift to a green economy. Mann reports that during the meeting, everyone emphasized the need for "ordinary" people to be involved in making decisions about transition. A green, just shift had to be "democratic" and "participatory," the participants avowed. Despite this insistence—or perhaps because of it—Mann found that "by the time of the day's closing plenary discussion, I could not shake the suspicion that confronting even the lower-intensity transitions was unlikely to involve 'democracy.'"[2] The main reason: the scale, speed, and intensity of change required to address the causes of climate change seemed to be unattainable via democratic practices. This seemed to Mann to be true not only through actually-existing democratic practices (i.e., voter-driven electoral politics) but also via models of deliberative democracy many of those at the CCPA event championed.

What might a politics appropriate to a transition, just or otherwise, look like? What issues are central to the processes and practices of constituting a politics of solarity? What questions should we pose to existing models of the political, including forms of radical politics, in light of the destructiveness of fossil capitalism and the promise of solarity? One possibility is a kind of solar liberalism whereby environmentalists and consumers act in the public sphere to pressure political representatives and institutions to accelerate transition to solarity by investing in the development of its infrastructures. Depending on your perspective, the politics of liberal

1. Geoff Mann, "Who's Afraid of Democracy?," *Capitalism Nature Socialism* 24, no. 1 (2013): 42–48.

2. Mann, 42.

solarity have either the advantage or the disadvantage of guaranteeing that social, political, and economic life will remain more or less unchanged, except for the energy source that powers it. Gregory Lynall reminds us that the history of solar power emits a "flexible technopolitics," flipping from right to left, from divine right to democracy and energy justice, in accordance with distributive power, with specific geographies and modes of scientific appropriation.[3]

A second possibility is a revolutionary transformation of fossil capitalism. In this scenario, the apparatus of the state is seized by the party of the organized underclass, and directed toward dismantling the inegalitarian and environmentally destructive infrastructures and relations of fossil capitalism and systematically replacing these with egalitarian and environmentally just infrastructures and relations of solarity. A "climate Leninism" of this sort would certainly be transformative in at least some respects.[4] However, there are strong historical reasons to be wary of the concentrated power of the sovereign state and its necropolitical, environmentally destructive effects, even in cases in which this power is exerted with socialist intention.[5] The case for the centralized authority of the state as the political instrument for achieving solarity is further complicated by the fact that, in the present conjuncture, governments that have shown themselves willing to exert the power of the state to enforce radical social and economic change tend to tilt to the right, not to the left.

A third possibility is a more decentralized, iterative, and combinatory politics that invests the possibility of solarity in democratically controlled local initiatives that develop and maintain sustainable, just, and responsible energy infrastructures. Such strategies are ori-

3. Gregory Lynall, *Imagining Solar Energy: The Power of the Sun in Literature, Science and Culture* (London: Bloomsbury, 2020), 8.

4. Kai Heron and Jodi Dean, "Revolution or Ruin," *E-Flux Journal* 110 (2020), https://www.e-flux.com/journal/110/335242/revolution-or-ruin/.

5. Thea Riofrancos, *Resource Radicals: From Petro-nationalism to Post-extractivism in Ecuador* (Durham, N.C.: Duke University Press, 2020).

SOLARITY IN TOGO
Claire Ravenscroft

This entry comes to you from a village in northern Togo, where a storm has knocked out the new and skeletal but otherwise reliable electricity grid. The proprietor of the local cyber café allows members of the community to use his solar-powered sockets to charge laptops and phones. He is called "le maire de Farendé" as a tribute to his fairness and generosity. Most people in Farendé farm, cook, and clean using solar power harvested the old-fashioned way, but the solar panels are a welcome source for those whose labor and entertainment draw on electricity. Whether or not the grid in Farendé is humming along, the cyber mayor offers power free of charge and on a shared basis with other members of the community.

Although conceivably produced by the very same factories using the very same minerals extracted from the very same mines, a solar panel in Farendé and a solar panel in Berkeley offer different lessons to our energy justice praxis. To be sure, neither is "absolved" of its production via capitalist arrangements of labor, resources, and power—like all commodities, solar panels are materializations of capitalist productive relations rather than their kryptonite. Yet, only the panels in Farendé permit us to identify the forms of solar power we want and help us to determine how to put them in place. They are a starting point, one among many, for a revolutionary political project of solarity.

ented simultaneously to energy transition and to a transformation of the social, political, and economic relations that subtend all energy formations. As autonomous enterprises, these cooperative, fugitive, and coalitional solarities present a paradox: a dramatic impact on the subjects and communities in which they arise and relatively modest impacts at systemic levels.[6] The future of this politics of solarity might depend on whether it tends in the direction of federation or isolation, and the institutional forms this tendency takes.

Dismantling fossil capitalism and achieving a solarity that is something other than fossil capitalism with sunny ways will re-

6. Jordan Kinder, "Solar Infrastructure as Media of Resistance; or, Indigenous Solarities against Settler Colonialism," *South Atlantic Quarterly* 120, no. 1 (2021): 63–76.

quire confronting the challenge of scale. Capitalist globalization and fossil fuels have generated a material condition in which the scale of our primary economic systems—extraction, agriculture and food, manufacturing, communication and transportation, services and finance, logistics and security—far exceeds the comprehension and jurisdiction of governments and their citizens.

Liberal political theory has no answer to scale. It falters and stammers in the face of it. Liberalism's core political tenets (autonomy, property, and security of the individual) and its chief political instrument (deregulated markets and private investment regimes) derive from a smaller time and a different set of political problems. Its impetus predated the scaling up of fossil fuel dependencies, the normalization of transnational corporations, and today's globalized population pressures. But we, unlike our predecessors, confront not simply the threat of the Machiavellian magistrate but rather new menacing hyperobjects (e.g., global warming, the global wealth gap, transnational racial capitalism) that resulted from choices made long ago and that defy the old political calculi to solve them. To these, liberalism has no answer; it has no tools to respond.

Yet many of us hang on to the little gestures of agency that liberalism offers. In the limited field in which we operate as individuals, solar praxes do indeed proliferate today: locavorism, PV rooftops, organic markets, veganism, composting, the biking commute. These small acts against things as they are give the late liberal subject a small voice of protest against fossil capital and the second nature we inhabit, an opportunity to gesture toward an imagined life after oil. And as long as we live in this political body, we will believe that each pull of a product off the grocery store shelf and each morning commute can send a meaningful signal, a little shiver, off to markets, to neighbors, and to policy makers about the content of our values. Mystified, we will yet hang on to the hope that we nudge Earth toward sustainability.

Solarity in the age of global warming (if it is to carry forward the project of democracy) requires something different, something to meet scale with scale, something to level this uneven playing field that is so overdetermined by a centuries'-long alliance between aggregated wealth and the fossil fuel industries. In other words, for solarity to punch in the same weight class, it must scale up. We all know what that means, that it can only mean one thing: solidarity. To get there, to inch toward solarity, means purging the libertarian fantasy that each of us matters prior to and apart from our relations—that there is no "we," that our relations do not matter. Solarity means asserting that we matter only by relation.

Storytelling and Worldmaking

THE TURNS TOWARD SOLARITIES, present and future, are not solely material turns. This transition requires new stories, myths, and forms of (en)visioning and imagining the world. Stories and myths are tools of immense possibility that provide powerful means of creating different worlds and making new futures, and of seeing the present in new ways. The stories we tell of solar pasts, presents, and futures are crucial means of unsettling received knowledge, reviving forgotten and occluded histories, and building future worlds. As Ruha Benjamin put it in a recent lecture, "imagination is a contested field of action."[1] Imagination, story, and art are inherently political; they are integral to material and experiential life and hold lessons and tools for building new worlds. Here we present a few examples of artistic, fictional, and speculative solarian worlds as prompts and examples of the imaginative possibilities of solarity.

Olafur Eliasson's installation *The Weather Project* opened in the Tate Modern's Turbine Hall in 2003. As its title suggests, the installation was an attempt to explore the experience of the weather by bringing it into the museum. The installation reproduced the mists and clouds of London's streets in Turbine Hall to give visitors a chance to reflect on the ways in which cities mediate their

1. Ruha Benjamin, "Race to the Future? Reimagining the Default Settings of Technology and Society," Mossman Lecture, McGill University, Montreal, October 28, 2020.

experiences of weather. However, this aim of the project was likely lost on most people who came to the Tate. The real attraction was the giant, bright orange-yellow sphere that Eliasson placed near the ceiling, and perhaps, too, the huge mirror on the ceiling that reflected everything back at viewers. It was the sun that came into the gallery, and there was no doubt that this is what everyone came to see. Visitors sprawled on the ground, turning the floor of Turbine Hall into a Mediterranean beach. In the rays of hundreds of monofrequency lamps, they came inside seeking warmth and light and to commune with a sun that they rarely found present with such intensity in the skies above London. *The Weather Project* is a misnomer for this installation. A better name might be "Solarity."

Reviewers and critics of Eliasson's faux sun pointed to the ways in which it worked to rewire actions and expectations, cutting through the apparent rationality of a busy London workday and providing those who dropped in with "new kinds of engagement with a world fraught with social and environmental concerns."[2] Eliasson sees it as "a subject that implied 'community' and that was open-ended. Predicting weather is one way we collectively try to avoid the unforeseeable, which our lives are always about. The weather is a subject about which a community may also permit a high degree of disagreement: I can say 'I hate the rain,' you say, 'I love it,' and you may still think I am a nice guy."[3] Many critics drew connections to the sublime or to sun worshipping and pointed to Eliasson's implied critique of modernity via the weather: it is now the only place in which city-dwelling humanity ever encounters anything like "nature." Not all are positive about the solar experiment carried out in *The Weather Project*. Louise Hornby points out that the installation's focus on "an ecology of individual encounter and feeling situate[s] the experiencing subject at the center, providing an analogue to

2. Helene Frichot, "Olafur Eliasson and the Circulation of Affects and Percepts, in Conversation," *Architectural Design* 78, no. 3 (2008): 30–35.

3. Olafur Eliasson, quoted in Michael Kimmelman, "The Sun Sets at the Tate Modern," *New York Times,* March 21, 2004.

the human centering that marks the era of the Anthropocene."[4] Hornby notes that the sun in the Tate actually offered no warmth and that the subjects lying on the ground together were interested in looking at their reflections on the roof—hardly the beginnings of a collectivity organized in relation to the challenges and promises of the solar. Complicating matters further, the piece had been installed in a setting that, at the time, was sponsored by BP, one of the world's largest fossil fuel companies.[5]

The process of developing a relation to the sun and its energy will involve missteps as much as steps forward. To make it work at all, we need to be humble, relinquishing the desire for mastery and perfection. We need to be alert to the fact that we can get it wrong (because we have gotten it wrong) and that we need the collective ideas and insights of people willing to share their diverse knowledges and experiences of plural solarities. In particular, we need to make way for what Macarena Gómez-Barris calls "submerged perspectives,"[6] which are described as emerging in and from "realms of differently organized reality that are linked to, yet move outside of, colonial boundaries . . . submerged perspectives that perceive local terrains as sources of knowledge, vitality and livability." Such perspectives "reveal a differently perceivable world, an intangible space of emergence, where rivers converge into the flow and muck of life otherwise."[7] To perceive solarity otherwise will require that we seek out, care for, and listen and respond to submerged perspectives wherever we might encounter them.

These perspectives are more nearby than we might think. "Solarity: After Oil School" took place on the traditional and unceded territories of the Kanien'kehá:ka people of the Haudenosaunee

4. Louise Hornby, "Appropriating the Weather: Olafur Eliasson and Climate Control," *Environmental Humanities* 9, no. 1 (2017): 60.

5. Gavin Grindon, "This Exhibition Was Brought to You by Guns and Big Oil," *New York Times,* May 26, 2020.

6. Gómez-Barris, *Extractive Zone,* 1.

7. Gómez-Barris, xx.

nations. In the machinima titled *She Falls for Ages,*[8] Haudenosaunee Mohawk artist and scholar Skawennati tells the story of Sky World, a "faraway place" whose people have "harnessed geothermal, wind and solar power and are brilliant botanists." As the narrator recounts, "one of their greatest creations is the Celestial Tree. Developed over thousands of years of careful cultivation, the tree's blossoms emit light. In fact, they light the whole world." The story's central figure is a telepath named Otsitsakaion. As Skawennati explains, "when she learns that her world is dying, she knows what must be done; she must become the seed of the new world."[9] Sky World is a faraway place, but the work that must be done to reach it starts right here.

Stories like this help us understand what must be done, and what must be undone, in response to a world that is dying. They give us the gift of feeling and thinking otherwise. The advent of solar energy has been treated by many as a wondrous silver bullet, promising a resolution to environmental trauma that will leave much else as it is: soon enough, we will all have ample energy and the powers that come with it, *and* it will be clean! If only it were so simple. The truth is, "one can still only imagine a world in which seven billion people had equal access to free power and could thereby take hold of their inborn solarity."[10] Innumerable desires are wrapped up in our understanding of the sun and its energies and innumerable historical and material obstacles standing in the way of those desires. These desires extend from hopes that we might adopt radically different ways of being in relation to one another and to the nonhuman others with whom we share the planet, to fantasies of powering extractivist capitalism on the cheap. Solarity is thus a space of ethical indeterminacy and political struggle, a structure of

8. Skawennati, "She Falls for Ages," Aboriginal Territories in Cyberspace/Obx Labs, 2017, http://www.skawennati.com/SheFallsForAges/.
9. Skawennati.
10. Boetzkes, "Solar," 317.

desire in which energy, climate, and attachments to infrastructure converge in contested spaces of imagined and material transition. As Rhys Williams observes in the context of solarpunk science fiction, "if energy transition is to be a battle of hearts and minds as much as PV panels and lithium batteries, a serious engagement with energy imaginaries is the means to understanding and marshalling them."[11] The problems *and* opportunities that might develop as a result of the advent of a solar world need to be at the center of our questioning and our struggles. These are seeds of new worlds that we must cultivate with care, and this work is ongoing. It is the work of the time we will spend together.

In envisioning a possible framework for solarity that might be inclined to support solidarity and environmental responsibility, we might also look beyond ideals of work and labor toward games and play. How do we build a revolutionary solar system that galvanizes multispecies communities in the spirit of sympoetic worldmaking?[12] Additionally, how do we conceive of revolution and energy outside of their traditional valences in an effort to build a just solar system—one predicated on the principles of environmental justice?

If the promise of solarity is a promise of better relations between different humans and nonhumans, it must be accompanied by stories, arts, tools, and crafts that celebrate and sustain collective flourishing. Our attention to nonhuman and non-Western kin reveals that many such stories already proliferate: coral, algae, trees, and other sun worshippers already tell us about how to build worlds otherwise. But we must learn to listen. Making new worlds will necessarily involve mistakes, but approaching such a project with humility and a willingness to be wrong, to learn from our solarian comrades across species and cultures, will help us make fewer and less egregious mistakes. In this contemporary moment of cascad-

11. Williams, "This Shining Confluence," 20.

12. For a discussion of sympoiesis—a term used to refer to a form of collective productivity via the work of multiple and multispecies actors—see Haraway, *Staying with the Trouble*.

ing planetary crises, we require more responsive and responsible arts of storytelling and worldmaking. We require stories that move away from solitary, individual heroes to multispecies stories that are grown over time, stories that are intertwined with other beings and celebrate not individual feats but the mutual creation of new ecosystems. To thrive collectively requires listening, learning, and making collectively.

Fumbling toward Solarity

SOLARITY IS NEITHER UTOPIAN NOR DYSTOPIAN, neither bright nor bleak, but a mixture of the two tendencies. If we take solarity in its utopian aspect, it becomes a horizon against which we can measure the present, evaluate our own actions. None of us are bright (utopian) solar subjects. Solarity is not an achieved state but a potential horizon against which we might evaluate the present, and evaluate ourselves.

The only relationship to solarity of which we are capable is one of imperfection and insecurity against an ideal horizon. The achievement of solarity demands radical changes to infrastructures, social relations, values, habits—to the human-made world in which we live, the ways in which we live in it, and the relations between the two. As such, we are thrown into a position of uncertainty, attempting to act, speak, make, live, against the grain of our own experience and learned behavior. In attempting to make a different world, we are thrown into the position of inexperience.

We don't know yet how to live in solarity; we are not, and don't know how to be, solar subjects. The world that would produce such a subject has not (yet) come into being. That leaves us in a position of radical insecurity, where our knowledges, embedded in practice and habit and tradition, concretized in our infrastructures, articulated in our politics, are insufficient, improper, polluted, and polluting. Solarity confronts us with the challenge of learning to live and do and think and love again, to hold ourselves askance to the world we knew, to estrange ourselves from both the world we know and ourselves, to begin to be differently and to build differently.

A key difficulty here is the willed relinquishing of expertise. We have learned how to successfully be in the world. But this experience is unfit for solarity, and the latter will not come into being if the former is not relinquished. We are thus faced with the demand to become children again, stripped of the limiting confidence of habit. The proper attitude toward solarity in the present is insecurity—to acknowledge that we lack expertise, that we do not know how or what to do. And yet we have to do something, so we act insecurely, we act precisely against what we know, to create space for new and other knowledges. Solarity recasts fossil-fueled adulthood as a mistake, as a foolish and dangerous mode of being and set of knowledges, despite its great confidence in itself. The proper attitude of insecurity is difficult to dwell in as an adult, but we might turn to the toddler as an emblem for proper solar subjectivity in the present—to try and fail and try again with no failure of confidence but rather a joy in the simple fact of doing better, bit by bit.

There is no place in the present for a finished statement of solarity, only provisional and imperfect gestures and efforts and figures. This holds for any and all areas of practice: What might solar art look like? Solar politics? Solar community? Solar education? We strive to estrange the present in new practices, and what we produce now from our fossil-saturated circumstances is to the horizon of bright solarity as the crude but joyful scrawl of a child is to a masterpiece. Or to be more exact, within the most sophisticated and radical products of the present, we will discern, at best, crude traces of that alternative future. It is possible that this crudeness is also a necessary quality of solarity, a quality of imperfection, of openness to revision.

Solarity in the present is taking pride in a shitty first attempt, and the commitment of the second attempt, in contrast to the practiced motions through which flows the status quo. Solarity is fumbling toward something that we don't know and can't quite figure. Solarity

SOLARPUNKING

Laurence Miall

The punk and DIY aesthetic is maybe the best approximate form for the look and feel of early solarity. Punk was popular because it sounded like anyone could do it. And it sounded like real life. It didn't hide its blemishes—it reveled in them. If it is to succeed as something other than a glossy substitute for petroculture, solarity will have to be honest about its promise and limitations.

DIY is often associated with renovation. Like the amateur renovators they are, solarians will have to be prepared to be laughed at, get things wrong, get frustrated, get hurt, and get rejected. They will by necessity frequently be dirty and look silly. It takes an amateur far longer than it does a professional to build, repurpose, or renovate, but solarity will not be a quick fix. It will be trial and error. And we will have to get our hands dirty.

is discomfort and insecurity, because only then do we know that we are on unfamiliar ground. Solar community is shared frustration. It's the difficulty of acting together when the rules need making up as we go along, when goals and aims are unclear, when failure is encoded into success. None of us is an expert solar subject; we are all amateurs.

Solar panels have an efficiency of, at best, around 25 percent. The light strikes the panel, and excited electrons begin to bounce and jostle inside the material, but only some 25 percent of that activity finds its outlet. A quarter of the electrons initiate the wave of motion through the metal wires that is the electrical current; the rest simply don't connect. The path to solarity is difficult to find, and we have to accept a much lower rate of "production" than we are perhaps used to, and a much higher rate of play, of excitation that goes nowhere, the joy of bouncing around in the hope of striking the path.

Can we think of solarity as a somatic recognition, which is to say, you know it when you feel it? Like sunshine on skin, is it a condition that surrounds us? But can it also damage us should it be too intense?

We can only make the just solar future with compromised present materials. The process of transition is necessarily troubled, and the foundations of a solar future—the panels themselves—are a document of a utopian future that is also a document of barbarism. The question of urgency is here paramount. The globalized, fossil-fueled, capitalist mode of production stands ready as a coiled spring to produce a solar future, and the brief time frame provided by the shrinking carbon budget for 2C or 1.5C necessitates its deployment. Even if there were the will, there is not the time for the means of producing the solar future to be made commensurate to its utopian figure. Which is to ask—can we build a just future with compromised materials? Can solar technology, produced through the exploitation of labor and nature, power a utopia?

Infrastructure embodies and extends the relations that produce it, concretizing values, making durable desires, and facilitating certain ways of being while impeding others. The challenge of just solarity under conditions of climate urgency is to somehow develop within, through, and on top of that which it seeks to negate.

In finance, speculation is the practice of engaging in short-term, risky financial transactions. In workshops, movement building, and youth organizing, speculation is something different. It's a process, even a method, that taps in to the power of our collective imagination. Both imply risk, but one has the power to subvert the other. Speculative methods have the potential to breathe life into a conference, a classroom, even a movement. Making speculative art together offers an intimate way of building community, discovering shared affinities, permitting conversations about tough issues to unfold, slowly. Futurist speculative methods can help to reverse fear and anxiety and ultimately challenge paranoid capitalist logics that permeate disaster scenarios. And disaster and fear are rising with temperature records, raging wildfires, cities short of water, historic floods, and more. Fear is abundant, but so is our imagination. The possibilities of solarity find form through collective futurist imagining—when we begin to rethink how we grow our food, how we travel, how we work, how we communicate with each other, how we conceive of time, how we interact with the species and ecosystems on which we rely for survival, and how we

BUILDING A SOLAR PANEL

Jennifer Wenzel

I still remember the first time I tried to build a solar panel. My community and I (a group of fossil fuel subjects trying to be solar subjects) waited for the power to come on so we could see what to do. Am I doing it? Am I doing it right? It turns out that using more solder makes the work go better. Trying to listen to the explanation of the physics of the cell, how electrons balance each other out, 5 v. 3 = 4—that's all I remember, because making is hard, and who has time (mind?) to listen or think? But the adjectives tossed into the circle of work tables—"dirty," "lumpy"—building a chain of meaning as we use heat to link metal to metal, lexical connections to describe work that is not like writing at all.

build the trust, the participatory values, and the skills necessary to overcome tremendous odds.

The difficulty of just solarity, as with any utopian horizon, is the difficulty of its translation into positive images, words, representations. The utopian process is one of continual failure, and not so much moving toward as away from. This is the necessity of a confident yet insecure subject, and community. Of acting in the unknown. Of a leap of faith, of the negation of what is and the affirmation of what is not, of what cannot be known but only intuited from its universal quality—that which unites rather than divides. Conditions of climate emergency and climate urgency provide a new and difficult context for utopian thought, as it demands positive and manifest action. How to act in uncertainty, yet also act positively? How to aim at utopia, but also knowingly deploy and draw on and so propagate the productive capacities of the present to maintain the boundary conditions of the planet such that utopia may be possible within it?

The climate emergency situates utopianism with a new problem—the need to maintain the conditions of its own possibility not only within human thought and practice but in terms of a livable ecosystem.

There is a lot of concern at present to imagine alternative futures, to produce radical imaginaries that critique the present and figure alternative ways of living. The effort to grapple with solarities is one example of this. But there is now a concern with overproduction of futures, too many futures created, consumed, and discarded, going nowhere. The future not as colonized by the present but as negated through becoming transformed into futures as commodities. This overproduction is antagonistic to solarity. Solarity requires, instead, stories that take root. Stories that are not statements but essays, tries, gestures, that are returned to and elaborated, time and again. Stories that try and fail and try again. Stories to which we commit and which we allow to change the way we live now, which in turn reveals the limitations and failures of the stories and gives us the means to rework them.

Acknowledgments

In May 2019, a group of international scholars, students, artists, activists, and practitioners met at the Canadian Centre for Architecture in Montreal, Quebec, for "Solarity: After Oil School," three days of intensive reflection and collaboration on the challenges and possibilities of a social transition to energy systems and communities organized around the energy of the sun. Workshops, presentations, and discussions were held on themes including feminist solarities, Indigenous solarities, community renewable energy, revolutionary solarities, Solarpunk, and speculative solarities. Participants were invited to contribute subsequent reflections on these and other themes. These contributions became this collectively authored text.

We met on the unceded traditional territory of the Haudenosaunee and Anishinaabeg peoples, near the Kanien'kehá:ka Nation communities at Kahnawá:ke, Kanehsatà:ke, and Akwesasne. Kawenote Teiontiakon is a documented Kanien'kéha name for the Island of Montreal; the city is known as Tiohti:áke in Kanien'kéha and as Mooniyang in Anishinaabemowin. We are grateful to the long-standing custodians of the lands and waters on which we met, and we hope that they are honored by the work we have done.

We are also grateful to several people and organizations that supported this work, including the Petrocultures Research Group, the Grierson Research Group, Future Energy Systems at the University of Alberta, the Social Sciences and Humanities

Research Council of Canada, the Initiative for Indigenous Futures at Concordia University, the Center for Energy and Environmental Research in the Human Sciences at Rice University, Just Powers at the University of Alberta, and the University Research Chair in Communication Arts at the University of Waterloo. Special thanks to Lev Bratishenko, Fannie Gadouas, Kim Förster, Rafico Ruiz, Jordan Kinder, Mark Simpson, Imre Szeman, Sheena Wilson, and Leah Pennywark.

Further Reading

Bakke, Gretchen. "Crude Thinking." In *The Rhetoric of Oil in the Twenty-First Century: Government, Corporate and Activist Discourses,* edited by Heather Graves and David Edward Beard, 34–55. London: Routledge, 2019.

Barca, Stefania. *Forces of Reproduction: Notes for a Counter-hegemonic Anthropocene*. Cambridge: Cambridge University Press, 2020.

Bennett, Jane. "The Solar Judgment of Walt Whitman." In *A Political Companion to Walt Whitman,* edited by John Seery, 131–46. Lexington: University of Kentucky Press, 2011.

Berlant, Lauren. "The Commons: Infrastructures for Troubling Times." *Environment and Planning D* 34, no. 3 (2016): 393–419.

Bird Rose, Deborah. "Shimmer: When All You Love Is Being Trashed." In *Arts of Living on a Damaged Planet: Ghosts and Monsters of the Anthropocene,* edited by Anna Lowenhaupt Tsing, Heather Anne Swanson, Elaine Gan, and Nils Bubandt, G51–63. Minneapolis: University of Minnesota Press, 2017.

Brown, Brené. "Shame v. Guilt." 2015. https://brenebrown.com/articles/2013/01/15/shame-v-guilt/.

Clark, Nigel, and Kathryn Yusoff. "Queer Fire: Ecology, Combustion and Pyrosexual Desire." *Feminist Review* 118, no. 1 (2018): 7.

Cowen, Deborah. "Infrastructures of Empire and Resistance." *Verso* (blog), January 25, 2017. https://www.versobooks.com/blogs/3067-infrastructures-of-empire-and-resistance.

Cubitt, Sean. "The Sound of Sunlight." *Screen* 51 (2010): 118–28.

Cuboniks, Laboria. *Xenofeminism: A Politics for Alienation*. London: Verso, 2015.

Dardot, Pierre, and Christian Laval. *Common: On Revolution in the Twenty-First Century*. London: Bloomsbury, 2005.

de Onís, Catalina. "Energy Colonialism Powers the Ongoing Unnatural Disaster in Puerto Rico." *Frontiers in Communication,* January 29, 2018.

Fournier, Valérie. "Utopianism and the Cultivation of Possibilities: Grassroots Movements of Hope." *Sociological Review* 40, no. 1 (2002): 189–216.

Groys, Boris. "Georges Bataille: The Potlatch with the Sun." In *Under Suspicion: A Phenomenology of Media,* 115–28. New York: Columbia University Press, 2012.

Haraway, Donna. *Staying with the Trouble: Making Kin in the Chthulucene.* Durham, N.C.: Duke University Press, 2016.

Hardt, Michael, and Antonio Negri. *Multitude: War and Democracy in the Age of Empire.* New York: Penguin, 2005.

Heglar, Mary Annaise. "I Work in the Environmental Movement. I Don't Care If You Recycle." *Vox,* May 28, 2019. https://www.vox.com/the-highlight/2019/5/28/18629833/climate-change-2019-green-new-deal.

Ignatov, Anatoli. "The Earth as a Gift-Giving Ancestor: Nietzsche's Perspectivism and African Animism." *Political Theory* 45, no. 1 (2017): 52–75.

Kallis, Giorgos. *Limits: Why Malthus Was Wrong and Why Environmentalists Should Care.* Stanford, Calif.: Stanford University Press, 2019.

Kumar, Ankit. "Cultures of Lights." *Geoforum* 65 (2015): 59–68.

Levitas, Ruth. *Utopia as Method: The Imaginary Reconstruction of Society.* New York: Palgrave Macmillan, 2013.

Marx, Karl. "Introduction: Production, Consumption, Distribution, Exchange (Circulation)." In *Grundrisse,* 83–111. London: Penguin, 1993.

Seeds of Good Anthropocenes. https://goodanthropocenes.net/.

Stengers, Isabelle. *Cosmopolitics I.* Minneapolis: University of Minnesota Press, 2010.

Szeman, Imre, and Darin Barney, eds. "Solarity." Special issue. *South Atlantic Quarterly* 120, no. 1 (2021).

Todd, Zoe. "Indigenizing the Anthropocene." In *Art in the Anthropocene: Encounters among Aesthetics, Politics, Environments and Epistemologies,* edited by Heather Davis and Etienne Turpin, 241–54. New York: Open Humanities, 2015.

Toscano, Alberto. "The Disparate: Ontology and Politics in Simondon." *Pli: The Warwick Journal of Philosophy* 23 (2012): 107–17.

Tsing, Anna. *The Mushroom at the End of the World: On the Possibility of Life in Capitalist Ruins.* Princeton, N.J.: Princeton University Press, 2015.

Vansintjan, Aaron. "Public Abundance Is the Secret to the Green New Deal." *Green European Journal,* May 27, 2020. https://www.greeneuropeanjournal.eu/public-abundance-is-the-secret-to-the-green-new-deal/.

Wilson, Sheena. "Energy Imaginaries: Feminist and Decolonial Futures." *Mediations* 31, no. 2 (2018): 377–412.

(Continued from page iii)

Forerunners: Ideas First

Kenneth J. Saltman
The Swindle of Innovative Educational Finance

Ginger Nolan
The Neocolonialism of the Global Village

Joanna Zylinska
The End of Man: A Feminist Counterapocalypse

Robert Rosenberger
Callous Objects: Designs against the Homeless

William E. Connolly
Aspirational Fascism: The Struggle for Multifaceted Democracy under Trumpism

Chuck Rybak
UW Struggle: When a State Attacks Its University

Clare Birchall
Shareveillance: The Dangers of Openly Sharing and Covertly Collecting Data

la paperson
A Third University Is Possible

Kelly Oliver
Carceral Humanitarianism: Logics of Refugee Detention

P. David Marshall
The Celebrity Persona Pandemic

Davide Panagia
Ten Theses for an Aesthetics of Politics

David Golumbia
The Politics of Bitcoin: Software as Right-Wing Extremism

Sohail Daulatzai
Fifty Years of *The Battle of Algiers*: Past as Prologue

Gary Hall
The Uberfication of the University

Mark Jarzombek
Digital Stockholm Syndrome in the Post-Ontological Age

N. Adriana Knouf
How Noise Matters to Finance

Andrew Culp
Dark Deleuze

Akira Mizuta Lippit
Cinema without Reflection: Jacques Derrida's Echopoiesis and Narcissism Adrift

Sharon Sliwinski
Mandela's Dark Years: A Political Theory of Dreaming

Grant Farred
Martin Heidegger Saved My Life

Ian Bogost
The Geek's Chihuahua: Living with Apple

Shannon Mattern
Deep Mapping the Media City

Steven Shaviro
No Speed Limit: Three Essays on Accelerationism

Jussi Parikka
The Anthrobscene

Reinhold Martin
Mediators: Aesthetics, Politics, and the City

John Hartigan Jr.
Aesop's Anthropology: A Multispecies Approach

Ayesha Vemuri is a doctoral candidate in the Department of Art History and Communication Studies at McGill University.

Darin Barney is professor in the Department of Art History and Communication Studies at McGill University, where he holds the Grierson Chair in Communication Studies. He is the author of *Prometheus Wired: The Hope for Democracy in the Age of Network Technology*, *The Network Society*, and *Communication Technology* and coeditor of *The Participatory Condition in the Digital Age* (Minnesota, 2016).